OUR LIVING WORLD OF NATURE

The
Life
of the
Cave

Developed jointly with The World Book Encyclopedia

Produced with the cooperation of
The United States Department of the Interior

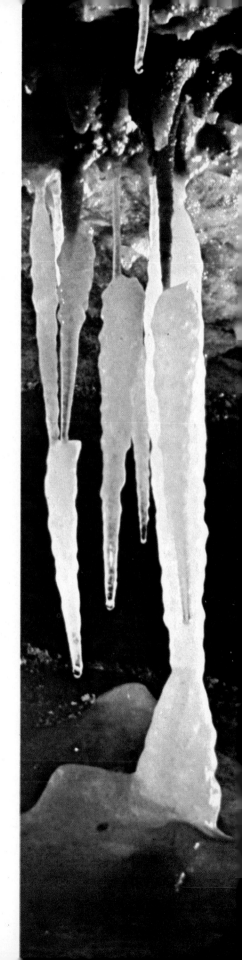

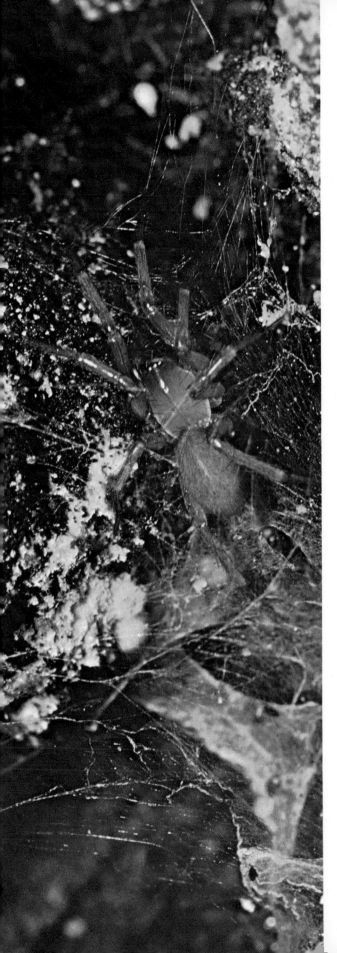

OUR LIVING WORLD OF NATURE

The Life of the Cave

CHARLES E. MOHR
and
THOMAS L. POULSON

Published in cooperation with
The World Book Encyclopedia

McGraw-Hill Book Company
NEW YORK   TORONTO   LONDON

CHARLES E. MOHR *is Executive Director of the Delaware Nature Education Center. From 1959 to 1966 he was Interpretive Specialist at the Kalamazoo Nature Center in Michigan, and from 1947 to 1959 he was Director of the National Audubon Society's educational center at Greenwich, Connecticut. He has been President of several organizations, including the National Speleological Society and the American Nature Study Society. In 1946 Mr. Mohr, a pioneer in the study of bat migration, led an expedition to study Mexican caves. Author of many technical and popular articles on natural history, Mr. Mohr was coeditor, with Howard N. Sloane, of* Celebrated American Caves *and a contributor to* Hunting with the Camera, *edited by Allan D. Cruickshank. As an Audubon Wildlife Tour Lecturer, he has participated in an educational program that spans the continent.*

THOMAS L. POULSON *has been associated since 1961 with Yale University, where he is an Assistant Professor of Biology. He is currently a director of the Cave Research Foundation and Executive Secretary of the Biology Section of the National Speleological Society. Dr. Poulson, whose doctoral dissertation concerned the biology of cave fishes, has published articles in several scientific journals. His studies of cave life have been concentrated in the eastern central states, but in the course of his research he has visited caves in other regions of the United States and in Mexico and Yugoslavia as well.*

Library of Congress Catalog Card Number: 66–24465

1234567890 NR 721069876

46003

OUR LIVING WORLD OF NATURE

Contents

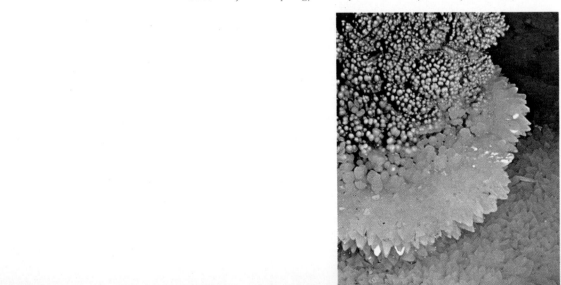

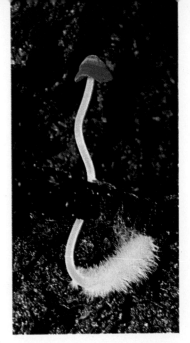

The World of

Darkness

It is a warm June afternoon in Kentucky, and the sun is shining brightly. Vireos and cardinals are singing in sun-laced branches of oak trees. A brook trickles along its limestone bed. Rough ledges on each side of the valley are green with mosses and lichens, and only here and there can you see the light gray rock underneath.

As you hike up the winding valley, these ledges grow higher until they become towering walls, nearly as tall as the trees that shade your path. For nearly fifty feet you have to walk in semidarkness under a natural stone bridge that spans the valley and blots out the sunlight.

Soon you are out in the sunshine again, where wild flowers bloom among the mosses and litter, and the leaves of the trees rustle in the wind. Continuing forward, you notice that the air is becoming cooler. Mosses and ferns grow thicker and deeper on the rock walls of the valley. Now the path takes a sharp turn to the right, and you round the bend to find that the valley is once again bridged over.

This time there is no daylight world just ahead. Peering into the gloom, you can see nothing distinctly. The pathway and the stream that it follows disappear into darkness. You have reached the entrance of a cave.

Headlamp, hard hat, sturdy clothing, and experience in mountain climbing equip a cave explorer for his descent into the unknown. Those who venture beneath the surface of the earth must come prepared for action and know how to use their equipment.

Slowly, cautiously, you take your first hesitant steps into the cave. Here and there water dripping from the cavern roof makes the path slippery. This monotonous dripping and the hollow gurgle of the stream flowing toward you out of the cave's interior are the only sounds to be heard. As you move along, it becomes more difficult to see ahead.

That's far enough for now. You're not prepared to go farther. You need lights. And even if you have brought flashlights or a lantern, you should not attempt to explore the cave without being sure that you have all the proper equipment. You might need ropes for climbing or an inflatable rubber boat.

What lies ahead?

What does lie ahead in such a cave? Must you climb over massive slabs of limestone where sections of the roof have fallen in? Will you have to lower yourself on ropes down deep shafts? Perhaps you will crawl hundreds of feet through low, muddy passages and wade across pools of cold water. Maybe you will walk for miles through vast echoing chambers with vaulted ceilings, or maybe you will find straight, level corridors stretching for great distances like a metropolitan subway system.

Surely explorers drawn by the lure of the unknown have visited this cave before you. Perhaps they have mastered the obstacles it offers, learned its secrets, and mapped the routes to be followed.

This might be one of those caves that contain hundreds of charred cane torches, left in centuries past by Indians whose desire for shining crystals overrode their dread of the darkness and unearthly quiet.

The first men to explore the cave may have found skeletons of prehistoric mammals trapped for eternity in a pit: bears bigger than our present-day species, or giant cats whose eight-inch fangs have earned them the name of sabertoothed tigers. Countless animals may have been caught in the dark crevices. The cave might be the grave of many an Indian who stumbled and fell to his death.

This spelunker is preparing to enter a deep pit. He has lashed his rope to a tree and has wound it around a metal spool tied to his waist. The spool will act as a brake with which he can control the rate of his descent.

Spelunkers explore an immense passage in Greenville Saltpeter Cave, West Virginia.

Like toothpaste oozing from a tube, the petals of gypsum flowers (*above*) grow longer as new crystals form at their bases. Jewel-like crystals of calcium carbonate (*below*) encrust a stalagmite rising from a pool of water.

Perhaps a romantic, half-forgotten story has prompted others to explore the cave that now awaits you. Was there some report of bandits hiding their treasure in this dark, solitary place? Did some escaped convict or some murderer with a price on his head make it his hideout? But most cave explorers do not need legends and fancies to spur them on, for they know that in these hollows under the earth real treasures are waiting for them. There is much to be seen that will amaze you and increase your understanding of the natural world. If you are an amateur geologist or biologist, you will find many things to puzzle over and study.

For the geologist, caves are among nature's unique workshops. In the one before you there may be vast galleries of *stalactites* and *stalagmites*, astonishing rock formations of great beauty and endless variety, ages in the making. You may find sparkling crystals, some breathtakingly delicate, others angular, some massive—objects so rare and strange that they could find a place in a great museum but are better left unspoiled for future generations to discover here, to wonder at, to enjoy.

The biologist may find this cave a wildlife sanctuary—a retreat for strange forms of life so specialized in structure and habit that they could not endure conditions on the earth's surface, where most other living things make their

Fragile crystals in the Caverns of Sonora, Texas, reveal a complex history of fluctuating water levels. Stalactites and stalagmites first formed when the room was dry. Later the room filled with water, and crystals of calcium carbonate grew like frost on the older formations. Now dry again, the chamber lures visitors who marvel at its delicate beauty.

homes. Here you might find odd creatures that emerge from the cave only in the blackness of night, and under only the right conditions of temperature and moisture. But other, still rarer inhabitants of the cave may never leave its absolute darkness—salamanders, fish, crayfish, spiders, and other small creatures whose whiteness seems out of place in a world of perpetual night.

Questions occur to you as you think of these curious animals that move without mishap down twisting inky-black corridors. How do they find their way? What makes them white? What do they eat and how do they find their food? To answer such questions, scientists have spent years studying cave dwellers; but many questions still remain to be answered by investigators willing to live close beside the animals and observe all the details of their daily lives.

Whatever lies within this cave—and the possibilities are almost unlimited—you will need flashlights or lanterns to illuminate your path and your surroundings. And you will

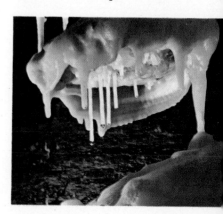

Hollow soda-straw stalactites form as dripping water deposits minerals at their tips.

Stalactites, rippled sheets of drapery, and a spangling
of grape formation decorate a passageway at
Carlsbad Caverns National Park in New Mexico.
The "grapes," each about a quarter of an inch in diameter,
grew on older formations when the room became
partially filled with water.

want a guide who has explored this cave and has firsthand
knowledge of its many features. Such help will enable you
to proceed confidently, observe intelligently, and understand why caves are so fascinating to *speleologists*, the scientists who study them.

A summer visit to the cave

Back at the cave entrance in early July, you are ready at
last to become a *spelunker*, a cave explorer. A local naturalist
who is familiar with many of the caves in this part of Kentucky has agreed to be your guide and has helped you
assemble the necessary equipment for the trip.

Before entering the cave, take time to record some observations at the threshold and examine the cave entrance
closely. No two entrances are alike, and the conditions at
each one vary with season, time of day, and weather. No
matter how often you visit a single cave, always try to take
notes on the environmental conditions and the plants and
animals that you observe. Note the air temperature now: it
is 90° Fahrenheit. The water is 58°. You will want to compare these observations with those you make when you come
back next winter.

The rays from the noonday sun behind you strike the left-
hand wall of the cave. Since the sun casts a shadow straight
north at noon, the cave entrance must face southwest. Your
compass confirms this reckoning.

The right-hand wall of the entrance is shaded and cool;
the sun will not reach there till late afternoon. Coming
closer to the entrance, you scare a phoebe from her nest on
a ledge. The naturalist tells you that he has met cows en-
joying the shade and coolness of a cave entrance on a summer afternoon, and that once he was nearly knocked down
by deer darting from the shadows. One spelunker friend, he
says, met a skunk in a narrow crawlway entrance. What sur-
prises may await a cave explorer in bear or mountain-lion
country!

Caves are not always horizontal
tunnels in hillsides. The entrance
to Fern Cave, Alabama, leads
to a shaft that plunges straight
down for 425 feet.

17

SCIENTISTS WHO STUDY THE LIFE OF THE CAVE

Each year more and more scientists
are descending beneath the surface of
the earth to study life in the cave
environment, the mysterious world of
"inner space." Equipped with everything from simple dip nets to complex
electronic devices, these biospeleologists
are seeking the answers to a variety
of questions: What special adaptations
permit cave animals to survive in a
world without night and day, without
green plants to provide food? What
events have isolated these animals in
caves? Do biological clocks allow them
to anticipate the spring floods that
will wash food into the cave? Why are
so many of them blind and white? By
studying how animals live together
in the drastically simplified environment
of the cave, biologists hope to gain a
better understanding of life in the
more complex and variable environments
on the surface of the earth.

*With compasses, measuring tapes,
and infinite patience, a party of
explorers maps a cave. Mapping
is an important first step in the
systematic study of a cave
and its life.*

A thermistor psychrometer, a pistol-like instrument, measures the moisture content of air in hard-to-reach pockets and crevices. The relative humidity of air sucked in through the slender nozzle is registered on a meter.

A sheet of tinfoil hanging from a pole is a simple but effective device for measuring airflow. The ruler on the floor serves as a gauge for indicating differences in velocity. Unless a cave has more than one entrance, air movement generally is slight.

These biologists are collecting tiny
aquatic animals from a shallow
pool. By using a tube similar to
an eyedropper, with a rubber
suction bulb at one end, they can
suck up the fragile specimens
intact.

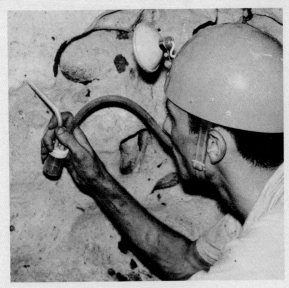

A biologist takes a tiny bug
from a dry wall by sucking it into
a collecting jar. Fine screening
across the rubber tube traps
the specimen in the jar—not in
the biologist's mouth!

Its tail twitching nervously, a phoebe eyes an intruder just inside the cave entrance. In caveless areas these agile flycatchers plaster their moss-and-mud nests to rock ledges, under bridges, and even on porches and in abandoned buildings.

The shadowed wall looks damper than the wall that gets more sun, and it is. More moss and ferns grow on it, but there are fewer wild flowers and other small broad-leaved plants here in the shade. If the cave entrance had faced north, you would not have found so much plant growth, since not enough light would be available. Or even if it faced south, any entrance much smaller than this one would cut off most of the light. A crawlway entrance, for example. would have hardly any green plants.

The twilight zone

Move slowly into the cave, giving your eyes a chance to adjust to the dim interior as you search for signs of life. The lacy ferns and clumps of moss that grew luxuriantly around the entrance are no longer as evident. But for a little way beyond the last recognizable plants, the cave walls and the

This dime-sized land snail normally creeps through the forest by night, but it sometimes wanders into caves. Its bulbous eyes protrude from the tips of slender stalks that can be lengthened or shortened at will.

ISOPOD

AMPHIPOD

Thousands of species of isopods and amphipods—tiny crustaceans related to shrimps and lobsters—live throughout the world. Isopods, with legs all more or less alike, are flattened from top to bottom. Amphipods, with several kinds of legs, have bodies flattened from side to side. Most cave-dwelling varieties are white and blind.

rocks jutting out of the stream still look green due to the presence of millions of microscopic green algae. The naturalist kneels down next to the stream flowing out of the cave and turns over a few stones. They are gray or brown underneath where the light does not reach; the green algae develop only in the presence of light.

Here in the stream, in the region where there is still some light, you find many little aquatic creatures living under the rocks. There are several immature insects housed in tubes of sand they constructed for themselves; they will mature into caddis flies, which are small mothlike insects, not really flies. You discover several brown flatworms with arrowlike heads and tiny crescent-shaped eyes, and you see some pale brown isopods, flattened horizontally, and a few shrimplike amphipods, flattened vertically. You move a large rock and uncover a brown crayfish, which, on sighting you, rears back and snaps its heavy pincerlike claws.

While you are kneeling beside the stream, take some more temperature readings: the water is 56°, the air 70°.

The pupils of your eyes are still expanding, and you can now see farther into the cave. After proceeding another twenty feet or so, pause to measure the dim light with your photographic light meter. The meter registers the light intensity almost at zero, indicating that its photo cell has reached the limit of its sensitivity. But your eyes are still sensitive. Behind your pupils, now fully open, your retinas continue to function; unlike photographic film, the human retina can increase its sensitivity as the light dwindles. So long as there is enough light to permit human vision, you are in the cave's *twilight zone.*

Besides having dimmer light than the entranceway, this zone has a more constant climate. It is always damp and cool, even when the weather outside is hot and dry. Hence the twilight zone is a natural sanctuary for many forest creatures. You can find some now if you look in the right places.

Turning your flashlight into pockets and recesses near the floor, you discover a salamander blinking at your intrusion. High on the wall a few daddy-longlegs, or harvestmen, back away from the light. You also make out mosquito-like midges hanging from the ceiling. But the most numerous animals are strikingly marked brown-and-yellow cave crickets with long legs and antennae.

From here on you'll need more light. Your carbide headlamp will throw a strong beam.

The variable-temperature zone

Now you have passed beyond the twilight zone. There is no light from the entrance. You could prove this complete absence of light by exposing a piece of ultrasensitive photographic film for a whole day and then developing it. The film would be blank and would produce a totally black print.

Your thermometer tells you that the air temperature is still dropping; it is now 63°. But the stream's temperature has not changed since your last reading; it is still 56°.

On the floor, well beyond the last of the green plants, are tufts of white and gray hairs, an inch or two in length. They remind you of mold on stale bread. And that is just what these hairs are: mold growing on the *scats*, or droppings, of a raccoon that wandered partway into the cave.

You catch sight of a bat hanging upside down. The naturalist guide says that this species, which he identifies as the "social bat," is one of the few in the eastern United States that make caves their summer home. It has a long string of aliases—Indiana bat, pink bat, cluster bat, and social bat—so

The colorful cave salamander is a common inhabitant of the twilight zone. The six-inch-long amphibian usually stays in damp cracks and pockets in the rock and ventures outside only when the air is moist, usually at night.

Cave crickets swarm across rocks in the twilight zone.

These tiny isopods are blind and colorless, like most animals from the dark interior of caves. Two kinds of supersensitive antennae guide them as they creep slowly across the bottoms of streams and pools in their endless search for food.

its scientific name, *Myotis sodalis*, is least confusing. (Unlike *Myotis sodalis*, however, many animals found in caves are so seldom seen that they have no common names at all. A guide for using their scientific names appears on page 202.)

What other kinds of animals do you see? On the ceiling are more crickets. They are pale brown and long-legged, and they have longer antennae and smaller eyes than those in the twilight zone. The first crickets you saw were *Ceuthophilus stygius*; these are *Hadenoecus subterraneus*. There are many more harvestmen on the walls here than in the twilight zone, but they appear to be the same kind. You also find a few tiny spiders and millipedes.

When you reach for a cricket, it leaps wildly, lands in the water, and skitters toward shore. Kneeling at the edge of the water to watch it, you see better swimmers: flatworms, amphipods, and isopods in full view, not hiding under stones as were the ones near the entrance. But these are white and eyeless.

Something much larger is walking on the floor of the stream from the deep water toward the shallow edge: a white crayfish, two inches long. It is a little smaller than the brown crayfish in the twilight zone, and more slender. It doesn't threaten with its claws or even appear to notice your presence until you stir the water. This creature is blind. You want to catch it and take it home, but the guide points out that the population of these animals is so small that you could disrupt the entire cave community by taking just one or two.

You are now aware of a puzzling difference between the animals in the water and those on land. Most of the terrestrial animals here in the variable-temperature zone have eyes and some coloring, but all those in the water are blind and white. Why should this be so? The guide explains that blind cave animals require a constant-temperature environment, and he says that at this point in the cave the stream temperature is a constant 56° throughout the year but the air temperature varies from season to season.

The constant-temperature zone

After walking several hundred feet farther down the passage, you stop again. Up to this point your thermometer has indicated that the air temperature has been sinking steadily. Now the air and water temperatures are the same: 56°; the air temperature will sink no further. The droplets of water

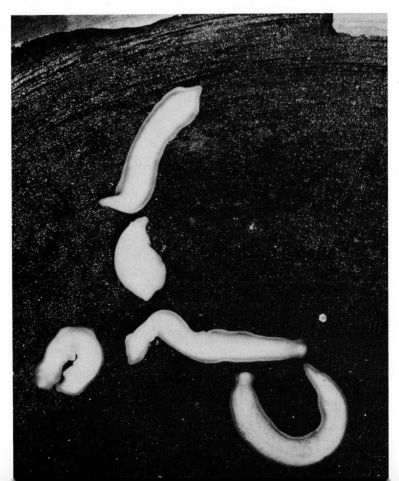

Ghostly half-inch-long flatworms are among the several varieties of cave animals that prey on isopods. As they glide across stream bottoms, they deposit trails of mucus. Some species may return later to feast on an occasional isopod trapped in the sticky secretion.

that have condensed on the walls indicate that the air here is saturated with moisture and will hold no more: the relative humidity is 100 per cent. Most of the drops have collected around tiny molds called *Actinomycetes*, which are responsible for the musty odor of the cave.

Now that the air temperature is constant, look for terrestrial animals again. There are still some harvestmen and millipedes on the walls, but not many. Looking more closely, you see that they are not the same as those in the variable-temperature zone: these are blind and have very little coloring matter, or *pigment*, in their outer bodies. Along the stream you see beetles about a quarter of an inch long scampering up and down the mudbanks. These, too, are blind.

In the constant-temperature zone you have a chance to study what conditions are best for the blind cave animals. You see that some parts of the stream have more food than others. Certain stretches are bare, containing only rock or sterile sand. But here and there you notice deposits of leaf and twig fragments, generally in silt-bottomed pools or in

A determined spelunker inches forward in a narrow passage. By-products given off by bacteria make the mud in caves particularly sticky and slimy.

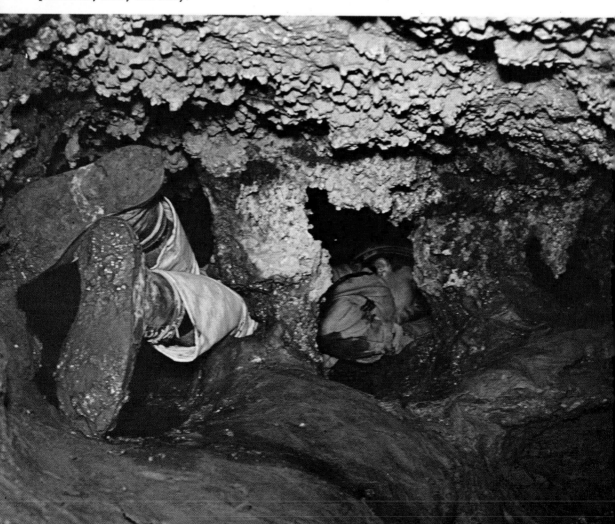

lagoons by-passed by the stronger currents. As you might suppose, in these places the isopods, amphipods, and crayfish are more common.

Since the passage ahead narrows and is nearly filled with water, you can best continue by following a diverging corridor up and away from the stream. You walk quite a distance before realizing that you have seen no animals. The ceiling becomes lower, and you must alternately crawl or duck-waddle. Occasionally you cross areas that seem very dry, but you raise no dust; the humidity must still be quite high.

At intervals there are sections of broad, high corridor where you can walk upright. In these places the corridor is as easy to follow as the stream passage you took into the cave. The walls up here have well-defined, regular scallop-like markings, which geologists recognize as evidence that a stream once flowed along this passage. The size and shape of the scallops indicate the direction and speed of the ancient stream.

The guide stops. Just ahead, he says, are dangerous vertical shafts, or *domepits*, into which you might fall. The first one rises thirty feet overhead and drops an equal distance

A sandstone cap rock forms the ceiling of a passage near the surface in Crystal Cave, Kentucky.

Seeping water carved this domepit by gradually dissolving the limestone.

31

At the edge of an abyss, an explorer gazes at a thread
of water that created the shaft plunging a hundred feet
or so beneath his perch. The jagged chunks of limestone
at the rim of the pit fell from the ceiling, probably
taking centuries to accumulate.

below you. Skirting it, you soon come to another shaft into
which you descend. The water dripping from the ceiling fifty
feet overhead splashes on the gravel-covered floor. This
water is clear; the gravel is clean, and there is not a single
animal to be seen. The water obviously carries no organic
material, no food.

You climb back to the passage above and continue past
more vertical shafts. The passage grows much larger now.
Indeed, the whole environment has changed. Here it is wetter and muddier, and once again you begin to see the tiny
blind brown beetles and the *Hadenoecus* crickets. But there
is no plant life, nothing that obviously serves as food. Where
do the crickets and beetles get their food? The answer, which
we shall discuss later, is a surprising one.

An oasis

You come to another vertical shaft. This one is partly collapsed. Water raining down its wall supplies a natural basin
in the rock, and a fine black coating covers the sand at the
bottom of the pool. The presence of flatworms and isopods
in the pool indicates that the black deposit must be edible—
it is organic—probably finely broken-down leaves. The guide
says that this passage is close to the surface. These odds and
ends of plant material wash in from the surface during spring
floods by way of small cracks and openings and make this
section of the cave an oasis for the inhabitants.

For the first time you see some stalactites and stalagmites.
The walls ahead are covered with *flowstone*, formations of
limestone deposited over thousands of years by water run-
ning down the walls.

Now you must climb a bit, because an enormous rock pile
—a *breakdown*—completely blocks the passage. The guide
says there is no way through this breakdown. It has sealed
off any extension that may have existed, and you will have to
turn back. But first see what animals you can find here.

The roots coming through cracks in the ceiling indicate
that you must be very close to the surface, but you cannot

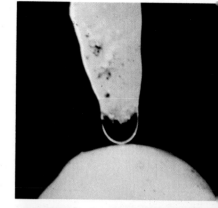

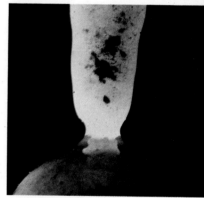

Seeping drop by drop,
mineral-laden water gradually
releases the calcium carbonate
that forms carrot-shaped
stalactites and dome-topped
stalagmites. These two pictures
of the same pair were taken
seven years apart and show how
stalactites and stalagmites often
unite into a column extending
from the floor to the ceiling
of a cave.

A massive breakdown clogs a passage with gigantic blocks of limestone.

detect any light. There are no beetles or millipedes. The few crickets you observe are *Ceuthophilus*, the twilight-zone variety.

Why should there be so few animals? The air temperature is still 56°, and the rock formations are dripping wet. However, there is a noticeable breeze, and apparently it is enough to discourage cave animals. The guide says that in the past he has found some animals by looking in sheltered places that are unaffected by the air currents.

He is right. There are crickets and harvestmen in ceiling crevices. Deep in a wall pocket you discover a blue-black salamander with tiny silvery spots. Its body nearly encircles a grapelike cluster of eggs. This is *Plethodon glutinosus*, the "slimy salamander." It is not exclusively a cave dweller, for it is often found in rocky woodlands. But this salamander has crept into the cave, an ideal place to lay its eggs. Now it will stand guard over them for more than two months in the cool dampness. The eggs are safer here in the cave than they would be if the salamander had laid them under a rock; there is much less danger of their drying out or being overheated.

Return to the entrance

Now you start back, moving as quickly as the low portions of the passage and your protesting knees will allow. After you have traveled through the dry upper passage and climbed down to the stream again, the guide says that you have been in the cave nine hours and at no time were you more than half a mile from the entrance. This is not surprising; time and distances are especially misleading underground. There are no familiar clues to guide you, and your sense of effort gives you no inkling of how much space you are covering or of how much time you have spent.

As you round the next bend approaching the entrance, you are puzzled by an eerie blue light. Getting closer, you see that the mysterious light is made by the full moon reflecting off the cave stream. Bats are darting low over the water in the entranceway, and you are surprised to see crickets head-

Found from New York to Florida and west to Missouri and Texas, the six-inch-long slimy salamander inhabits moist woodlands, sometimes laying its eggs in caves.

Tiers of flowstone and rippled drapery resemble a frozen waterfall. This complex rock formation resulted from the flow of mineral-saturated water down an uneven cave wall or a rockslide.

*Though nearly ready to hatch,
the big-eyed larvae still have
considerable yolk to use up.
The pea-sized eggs of the slimy
salamander are larger than those
produced by most salamanders
of the same size.*

SLIMY SALAMANDERS LAY THEIR EGGS ON LAND

Frogs and toads lay their eggs in water, and they hatch into plump, legless tadpoles.
Most of their relatives among the salamanders also lay their eggs in water;
their gilled, elongated, legged larvae swim in ponds or streams. Some transform into
land-dwelling adults, while others, including most of the blind salamanders, spend
their whole life in the water. The slimy salamander, however, lays its eggs on land.
A lover of damp, dark places, it goes so deep into rock piles that few naturalists
have ever uncovered the eggs. Only in the humid interiors of caves have the eggs
been laid where they may be observed, their long incubation period timed,
and the actual hatching witnessed and photographed. Within a couple of days
after hatching, the tiny salamanders scatter, heading for the cave entrance
and tunnel-like channels beneath the leaf litter of the forest, each fully independent
and ready for a career of hunting worms, insects, and other small prey.

*Encased in a mucous envelope,
about a dozen developing eggs
of the slimy salamander dangle
from the ceiling of a recess
in a cave wall. Possibly to protect
the eggs from predators,
the vigilant adult stands guard
for two months or so until
they hatch.*

*Newborn slimy salamanders remain
near the mother for a day or two,
then gradually wander off to fend
for themselves. Most amphibians
hatch in water and breathe
by means of external gills, later
transforming into land animals.
The slimy salamander goes through
this water stage inside the egg.*

ing out of the cave. Later on, when you learn more about these crickets, you will find that they come out to feed, but only on quiet, humid evenings.

This cave has a history

Before you depart from the cave entrance, the naturalist guide has a few more things to explain. He shows you a map of the cave which he has superimposed on a geological map of the area. When you see the cave in a wider setting, you understand why large volumes of water flow into the cave during heavy spring rains. A large watershed lies beyond the cave. Floodwaters follow normally dry channels and carry accumulations of woody debris and leaves into the cave. Water-transported organic matter provides most of the food used by many of the cave dwellers.

According to the map, a layer of sandstone caps the ridge. Ground water flows through the soil covering the sandstone, but it cannot sink directly down into the cave except through a few breaks in the sandstone roof. That is why you found such dry quarters and so little life when you first entered the upper passage; this part of the cave had a solid sandstone roof.

The vertical shafts you saw farther along and the pool oasis lie just outside the edge of the sandstone cap. Here water spilling off the sandstone can work its way downward through the soluble limestone, into the cave. For thousands

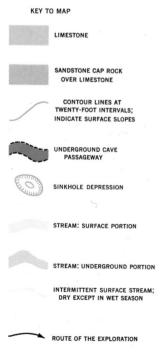

KEY TO MAP

LIMESTONE

SANDSTONE CAP ROCK
OVER LIMESTONE

CONTOUR LINES AT
TWENTY-FOOT INTERVALS;
INDICATE SURFACE SLOPES

UNDERGROUND CAVE
PASSAGEWAY

SINKHOLE DEPRESSION

STREAM: SURFACE PORTION

STREAM: UNDERGROUND PORTION

INTERMITTENT SURFACE STREAM;
DRY EXCEPT IN WET SEASON

ROUTE OF THE EXPLORATION

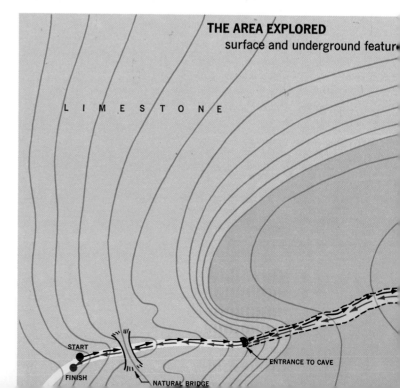

THE AREA EXPLORED
surface and underground featur[es]

L I M E S T O N E

START

FINISH

NATURAL BRIDGE

ENTRANCE TO CAVE

of years, with each heavy rain, water has descended through joints and crevices in the limestone and has gradually created these shafts. Long ago the oldest and largest of these caved in. During spring floods this is the place where leaves and organic matter come washing down to form the food-rich pools you noticed.

The geological map shows why the passage grew so much wetter as you went on. You had come very close to the edge of the valley, almost to the surface. You passed beyond the sandstone area to a region where there was nothing to prevent surface water from seeping down through the cracks and joints in the limestone. This was the location of the breakdown that made you turn back. Unquestionably, there are small openings here to the world outside. They admit the breeze which most terrestrial cave creatures avoid. The salamanders and crickets which you found here must be able to enter and leave the cave through tiny passageways in the breakdown.

A winter visit to the cave

It is a very cold morning in February when you return to make a second reconnaissance of the Kentucky cave. Your thermometer reads 20°. Walking along parallel to the stream, you notice that the water level is higher than it was during your earlier visit. It is easy to see that the stream is much warmer than the air, for mist is rising from the water.

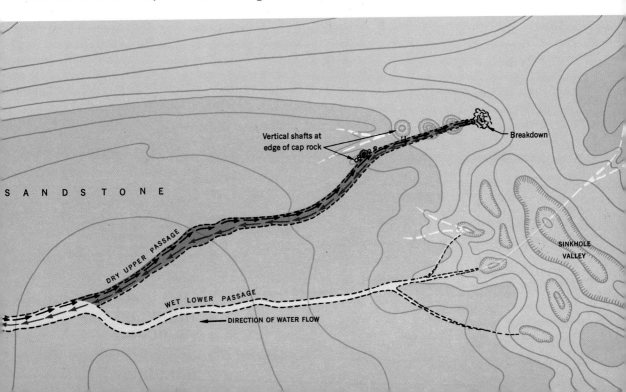

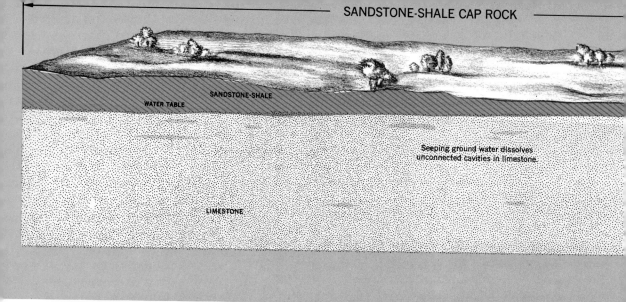

SANDSTONE-SHALE CAP ROCK

SANDSTONE-SHALE

WATER TABLE

Seeping ground water dissolves
unconnected cavities in limestone.

LIMESTONE

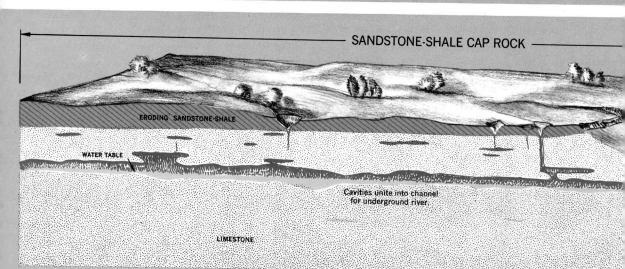

SANDSTONE-SHALE CAP ROCK

ERODING SANDSTONE-SHALE

WATER TABLE

Cavities unite into channel
for underground river.

LIMESTONE

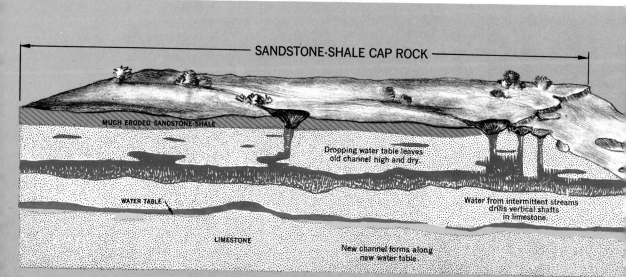

SANDSTONE-SHALE CAP ROCK

MUCH ERODED SANDSTONE-SHALE

Dropping water table leaves
old channel high and dry.

WATER TABLE

Water from intermittent streams
drills vertical shafts
in limestone.

LIMESTONE

New channel forms along
new water table.

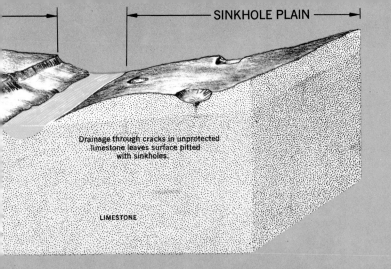

SINKHOLE PLAIN

Drainage through cracks in unprotected
limestone leaves surface pitted
with sinkholes.

LIMESTONE

HOW LIMESTONE CAVES
ARE FORMED

In the early stages of cave formation,
the water table lies close to the surface. Slightly acid water seeping
through cracks and crevices forms
small unconnected cavities by dissolving limestone in the saturated zone
below the water table.

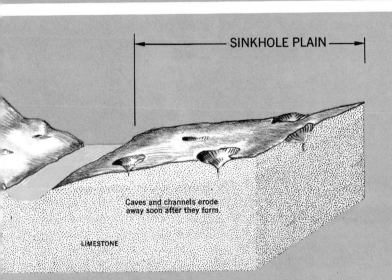

SINKHOLE PLAIN

Caves and channels erode
away soon after they form.

LIMESTONE

By the time the river has eroded
down to a new level, the cavities
have enlarged and joined into a continuous channel. Periodic floods continue to enlarge the openings. In the
area unprotected by the erosion-re-
sistant cap rock—the sinkhole plain
—sinkholes dot the surface where
rain water drains directly into the
limestone.

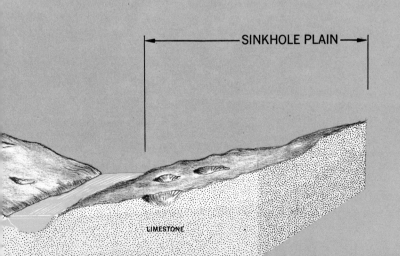

SINKHOLE PLAIN

LIMESTONE

Thousands of years later, the river
drops to a still lower level. The old
cave is left high and dry while another system of caves forms along
the new water table. Dripping water
deposits formations in the old cave.
Water draining through breaks in the
cap rock connects the old and new
cave systems by drilling vertical
shafts through the limestone.

SANDSTONE CAVES *such as the one in Canyon de Chelly National Monument, Arizona (right), form near the bases of sandstone cliffs. Flowing water or sandblasting by the wind gradually erodes cavities in areas where sand grains are weakly cemented. Many such shelters in the Southwest were used as building sites by prehistoric Indians. Few extend deeply enough to have a dark zone.*

SEA CAVES *such as Anemone Cave in Acadia National Park, Maine (above), are carved by the relentless pounding of surf against weak areas in the rocks of the steep coastal cliffs.*

NOT ALL CAVES ARE FORMED IN LIMESTONE

LAVA CAVES *such as Subway Cave in Lassen National Forest, California (below), are remnants of past volcanic activity. They result when a flowing stream of lava solidifies on the outside while molten lava continues to flow below the surface, finally leaving an empty tunnel. Several hundred are located in Lava Beds National Monument, California.*

When you round the bend and see the cave, you catch your breath. Winter has transformed this scene. The trees and bushes around the entrance are glittering with frost, and the fog has frozen into powdery crystals on everything it touched. Cold air flowing into the cave—you feel a tingling breeze at your back now—has frozen the water dripping from the ceiling.

Inside the cave a few slender spears hang from the ceiling. Below are many tall, flat-topped columns rising from the floor. These are the familiar stalactite and stalagmite formations that are regularly found in limestone caves; here they are made of transparent ice.

Small variations in temperature have brought about this display. Water seeping down from the surface is too warm to freeze until it reaches the cold air in the entranceway. Now the air converts it to gleaming ice formations. Most of the ground water drips to the floor before freezing, building translucent stalagmites.

It is so cold out here that you cannot stand still without shivering. In this weather the absence of terrestrial creatures is not surprising. Even if they could bear the climate, there is nothing for them to eat. Frost has killed the green plants you found during your earlier visit, or else they are dormant for the winter.

In the stream there are fresh pieces of rock with sharp edges and clean gray surfaces. They have been pried loose from the cave ceiling by alternate freezing and thawing at the entrance. Because of this yearly destruction the entrance area is wider and higher than the passage fifty or sixty feet inside and is the only area of the cave likely to be particularly hazardous. Thousands of years of weathering have driven the cave entrance back from its original site along the main river valley; now it occupies the head of a small valley.

The variable-temperature zone in winter

There is little light penetrating the cave on this gray day, and you will need lights almost from the very entrance if you are to see ahead and look for cave creatures.

Much as frost forms on cold windowpanes, excess moisture has coated every rock and twig at this cave entrance with glistening crystals of ice.

"It was like scaling a crevasse in a glacier," one visitor commented of his ascent from the glittering wonderland at Fossil Mountain Ice Cave.

Bundled up against the cold, summer visitors enjoy the beauty of nature's deep-freeze at Fossil Mountain Ice Cave. Stalagmites of ice (below) rise like huge teeth where dripping water has frozen on the floor of the cave. The frost on the ceiling condensed directly from moisture in the air.

WHERE SUMMER NEVER COMES

Temperatures at Fossil Mountain Ice Cave, Wyoming, are below freezing the year round. Since a cave's temperature approximates the average annual temperature aboveground, ice is fairly common in caves at high elevations in northern areas. Even at lower elevations ice may remain all year because a cave can act as a cold trap. In winter, cold air, being heavier than warm air, may settle to the cave's lowest level, where it cannot be displaced by lighter, warmer air in summer. With constant below-freezing temperatures, moisture in the air condenses into blankets of frost on rock surfaces, and water seeping in from the surface freezes into icy stalactites and stalagmites.

Nearly a hundred feet inside you find ice formations showing no signs of melting. All the way down the narrowing passage you have felt a cold breeze on your back. On this occasion, unlike last summer, the air is flowing in, not out: cold air, being heavier, flows under warm air into the lowest areas.

By now you have reached the region where, last summer, the constant-temperature zone began. But the cold winter air from outside has caused the constant-temperature zone to recede farther into the cave. The larger the cave entrance and passageway, the more cold air can enter. Some caves with small twisting passageways are little affected by the outside temperature.

At this point the air temperature is a chilly 38°, but your thermometer shows that the stream temperature is 56°, exactly the same temperature that it was last summer. You notice several isopods and a blind white crayfish.

With a little searching, you find clumps of harvestmen huddling together above you. A number of pink moths and

Hibernating harvestmen cover the wall of a sheltered cranny in the variable-temperature zone, safe from the frigid air outside. In summer the spiderlike creatures spread through the surrounding forest. Moths, crickets, bats, frogs, snakes, and many other animals escape the lethal cold of winter by entering caves.

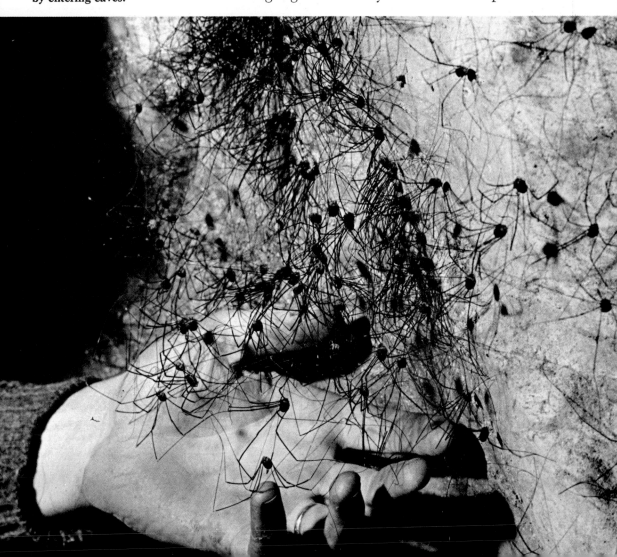

dozens of cave crickets are in residence on the ceiling. You find no blind beetles or white millipedes yet. Conditions are still too cold and variable for creatures accustomed to the eternal sameness of the constant-temperature zone.

All the occupants of the walls and ceiling in this zone—crickets, moths, harvestmen—are species that emerge from the cave in summer. For the duration of the winter they will remain dormant here in protected nooks and cracks, where the breeze does not reach them.

Your guide says that different species of bats show extraordinary ability to recognize differences in temperature and air movement, a sensitivity that has made them objects of extended scientific study. As you sweep your light across the ceiling, you sight a small cluster of *Eptesicus fuscus,* commonly called "big brown bats." Moving along, you see more of them, generally hanging singly or in small groups.

Bats are hard to please

Each species of bat has a remarkably precise set of conditions which must be met before it settles down for the winter. The big brown bat, for instance, hibernates only in relatively breezy locations where the temperature is 34° to 40°. Furthermore, it chooses locations where the humidity is at least 90 per cent. If the conditions change too much, the bat wakes up and moves to new quarters that meet its requirements. Speleologists can get a good idea of temperature, air movement, and humidity—the *microclimate* of a given place—just by observing what species of bats are hibernating there.

Farther down the passage you find more bats—gray, and somewhat smaller—which you recognize as the social bat. You recall seeing a few here in July. Now there is a tremendous colony. One dense compact mat covers a five- by six-foot area of the ceiling and contains possibly 5000 individuals. They are distributed with remarkable uniformity: a regular pattern of minuscule faces pokes out of the "bat rug." Speleologists have learned that each hibernating species seems to choose winter quarters as close to freezing as its constitution will allow. Big brown bats, as you saw, hibernate at 34° to 40°. These social bats apparently are comfortable at 36° to 45° where the relative humidity is very high.

Wide awake, a "lump-nosed bat" *(top)* **displays the basis for its other common name—the "long-eared bat." During hibernation it curls its ears in spirals and folds them down against its neck** *(bottom),* **possibly to reduce water loss from the delicate membrane. Bats carried from caves in winter may die of dehydration.**

At the next stopping place, a hundred feet farther down the passage, the temperature has risen to 48°, and the breeze at your back has diminished. Using a small instrument that he calls a *psychrometer*, the guide measures the humidity. It is 97 per cent.

Another variety of bat is hibernating here: *Myotis lucifugus*, the "light-avoider," commonly known as the "little brown bat." It looks much like the social bat, but instead of clustering in exposed dense mats, it chooses a depression or crevice for its dormitory. Here, where the temperature and humidity are just a bit higher than in the open and no air is stirring, the little brown bat fits snugly into every available hiding place. If you were to investigate the exact climatic conditions in these bat-frequented recesses, you would consistently get the same results: a temperature range of 43° to 50° and a relative humidity of at least 99 per cent. The almost imperceptible difference between this microclimate and that of the open ceiling is a matter of importance to the little brown bat.

Two ways to meet winter conditions

Some species of bats migrate to areas far to the south, where insects will be flying all winter. The ones you have just seen wait in caves, or other protected quarters, for the coming of spring and the return of insects to the night sky. The food they eat and the fat they accumulate before they become dormant must last them from fall until the following spring. With few breaks in their long sleep and with little shifting about, they can survive this long fasting period. In their deep hibernation, their breathing and heartbeat nearly stop, and their body temperature sinks to the temperature level of the air around them or of the rock to which they cling.

Since *metabolism* (the process by which all living things transform food into energy and living tissue) involves chemical reactions, cold slows it down. The colder the bats' winter quarters are, short of freezing, the longer their fat reserves will last. Each species seems to find the coldest environment it can tolerate.

Scientists do not yet understand the marvelous mechanism by which a bat is roused when the temperature drops dangerously close to freezing. Once a hibernating bat is disturbed by fluctuating temperatures, by the presence of cave

HIBERNATING SOCIAL BATS

explorers, or by the pressure of urine accumulated in its bladder, rousing begins and cannot be halted. The bat's heartbeat increases a hundredfold, and it shivers violently. As a result its metabolism quickly increases, and its temperature rises at a rate of nearly 2° per minute. Complete arousal of a dormant bat takes about ten to fifteen minutes, in contrast to the one to two hours required for it to enter a state of hibernation. Most bats cannot fly until their temperature reaches at least 85°.

When a bat is wakened by a drop in temperature, it quickly seeks a new dormitory. The social bat wakes up and shifts about more frequently than other species. It is such a supersensitive weatherman that it responds to temperature changes of a single degree.

The constant-temperature zone in winter

You know you have finally reached the area of constant temperature when you begin to see blind beetles, white spiders, and white millipedes, the permanent cave dwellers you found on your earlier visit. Like the bats they have shifted their living quarters. How far they move away from the entrance depends on how far cold air penetrates into the cave. Depending on the size of the entrance and passageways, and how much surface water may be carried in during floods, the location of the constant-temperature zone may be shifted drastically.

Now take a look at the quarters other bats have chosen.

There are none out in the open, and even the crevices are unoccupied. In one of the bowl-shaped pockets off in a far corner of the ceiling there is a single yellowish bat, a tiny furry ball. It is the smallest common bat, a pipistrel.

It takes only a moment, and a little discourtesy to the bat, to measure the conditions in its immediate surroundings, its *microhabitat*. Droplets on the bat's fur indicate that the relative humidity in this pocket is 100 per cent; your thermometer shows that the temperature is 56°, and there are no air currents. Since the microclimate here is much less variable than in the roosts of the three other bat species you found in the cave today, why don't you find more pipistrels? One reason is that the species is solitary rather than colonial in habit, and that a total cave population is generally numbered in scores, rarely in hundreds.

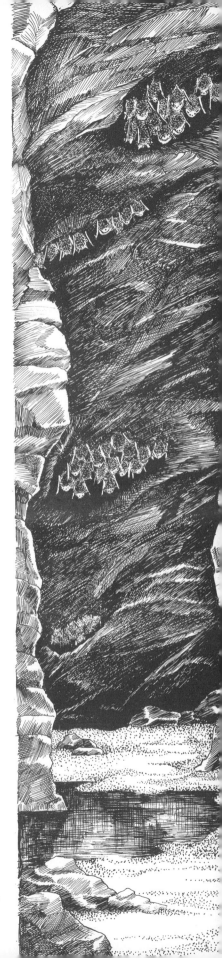

HIBERNATING LITTLE BROWN BATS

Entrance	Twilight Zone	Variable-temperature Zone		
AIR TEMPERATURE 90°	70°	67°	63°	58°
WATER TEMPERATURE 58°	56°	56°	56°	56°

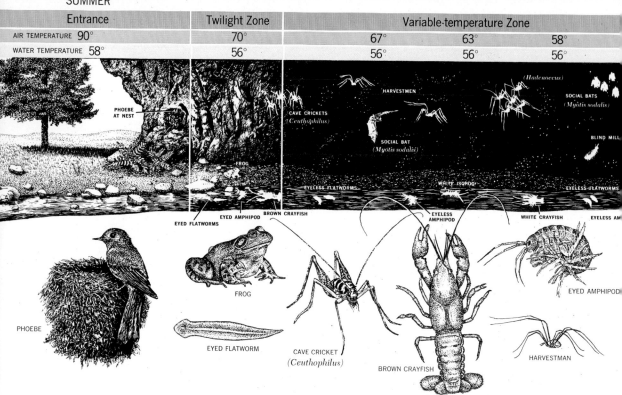

PHOEBE AT NEST

CAVE CRICKETS (*Ceuthophilus*)

HARVESTMEN

(*Hadenoecus*)

SOCIAL BATS (*Myotis sodalis*)

SOCIAL BAT (*Myotis sodalis*)

BLIND MILL

FROG

EYELESS FLATWORMS

WHITE ISOPOD

EYELESS FLATWORMS

EYED FLATWORMS

EYED AMPHIPOD

BROWN CRAYFISH

EYELESS AMPHIPOD

WHITE CRAYFISH

EYELESS AM

PHOEBE

FROG

EYED FLATWORM

CAVE CRICKET (*Ceuthophilus*)

BROWN CRAYFISH

EYED AMPHIPOD

HARVESTMAN

CAVE ZONATION, SUMMER AND WINTER

In a cave, as in many other habitats, variations in living conditions produce a series of life zones, with each zone supporting a distinctive community of plants and animals. The twilight zone extends from the entrance to the deepest point where light penetrates. It is the only area of the cave where green plants can grow. The variable-temperature zone is the area in which the temperature and humidity fluctuate with the changing weather conditions outside. It is colder in winter and warmer in summer, and it may increase or decrease in length. The temperature in the innermost area, the constant-temperature zone, is un-

Entrance	Twilight Zone	Variable-temperature Zone	
AIR TEMPERATURE 20°	24°	30°	38°
WATER TEMPERATURE 54°	55°	56°	56°

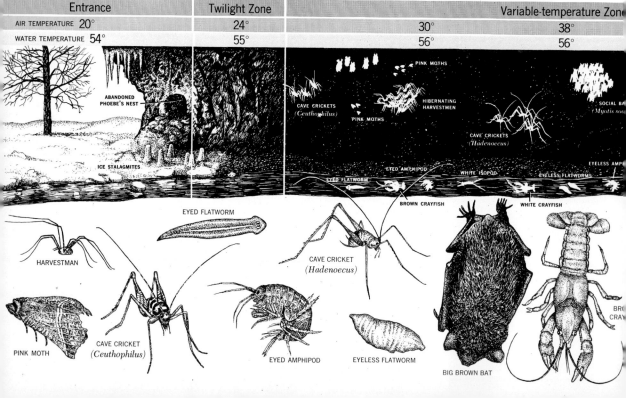

ABANDONED PHOEBE'S NEST

PINK MOTHS

CAVE CRICKETS (*Ceuthophilus*)

PINK MOTHS

HIBERNATING HARVESTMEN

CAVE CRICKETS (*Hadenoecus*)

SOCIAL BA (*Myotis suc*)

ICE STALAGMITES

EYED AMPHIPOD

WHITE ISOPOD

EYELESS AMP

EYED FLATWORM

EYELESS FLATWORMS

BROWN CRAYFISH

WHITE CRAYFISH

EYED FLATWORM

HARVESTMAN

CAVE CRICKET (*Hadenoecus*)

PINK MOTH

CAVE CRICKET (*Ceuthophilus*)

EYED AMPHIPOD

EYELESS FLATWORM

BIG BROWN BAT

BR CRAY

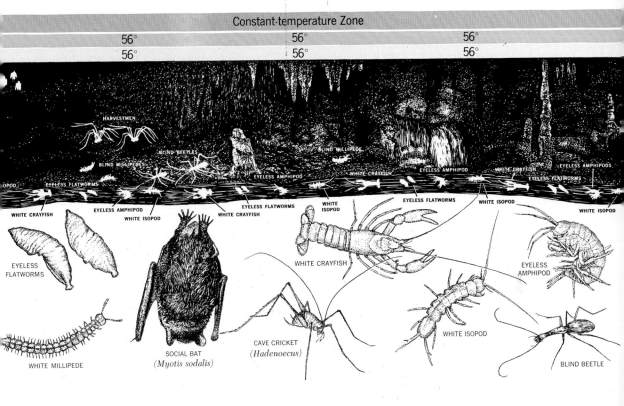

changing and approximates the average annual temperature aboveground. The beginning of the constant-temperature zone, however, normally is closer to the entrance in summer than in winter. Constant temperatures also occur closer to the entrance in water than on land, since water warms and cools more slowly than air. The animal population of each section of the cave changes from season to season as surface dwellers migrate in and out and as the full-time residents move back and forth, always staying within the zone to which they are adapted.

Constant-temperature Zone

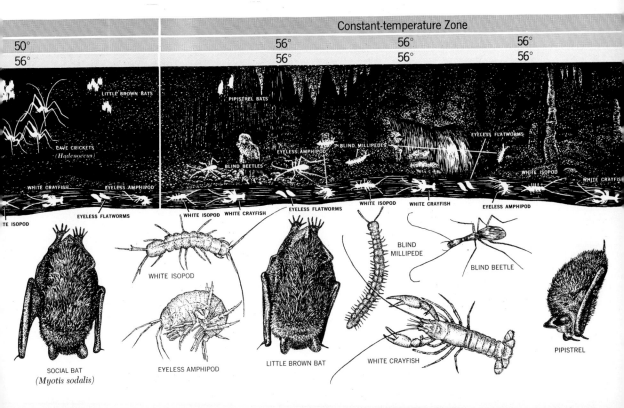

Zonation in the cave

As you have seen during this winter visit to the cave, each species of bat comes to rest in places where the temperature, air movement, and humidity precisely meet its requirements. The presence of a given species tells you, as accurately as instruments can, what the climate is in that spot. Indeed, you could probably locate the various zones in the cave without using instruments, merely by observing the positions of the hibernating bats. Each species is restricted to a special microclimate; some bats hibernate in the near-freezing temperatures relatively close to the entrance, while others hibernate in the more moderate temperatures and higher humidities farther inside.

But the bats are not the only living things in the cave to exhibit a striking *zonation*. On your summer visit you observed that the vegetation inside the cave changed dramatically as you proceeded farther from the entrance. The leafy plants were quickly replaced by mosses; eventually the mosses disappeared, and the algae on the rocks were the only green plants left. As you passed through the shadows of the twilight zone into the total darkness of the variable-temperature and constant-temperature zones, you saw no green plants at all, only molds and other fungi, whose air-borne spores settle and "feed" on organic matter.

Zonation occurs in every living community; it is a universal phenomenon in our natural world. You have probably noticed how climate and vegetation change with latitude as you travel north or south. And you may have observed how similar changes take place over a very short distance as you climb a mountain. But there are few habitats where the zonations of climate and of living things are as abrupt as they are in a cave. Within a few hundred feet you have seen fantastic changes due both to variations in temperature and humidity and to the lack of light. It is this lack of light in the cave which will concern us in the next chapter.

Beaded with moisture, a hibernating pipistrel clings with
sharp-nailed toes to a smooth cave wall. Unlike the
clustering social bats, the tiny "pigmy" bats hibernate
alone; few caves harbor more than several dozen
individuals. A full-grown pipistrel measures less
than two inches from nose to toe.

The Food *Pyramid*

It is hard to picture the cave environment beyond the twilight zone. You can see so little at a time, only the small portion illuminated by your gas lantern, carbide lights, or flashlights. Everywhere else that you explore, you are able to get some impression of what lies ahead: you see where you are going. Even an unclimbed mountain peak can be studied at a distance, in most weather. You can locate glaciers, timber line, and even the movement of mountain goats searching for edible plants.

On the prairie, your view takes in miles of tall waving grass; you recall the millions of bison that lived on this nutritious food as you watch their inheritors, the herds of domestic cattle, which are grazing and growing fat on it.

Along the coast you survey countless acres of bobbing seaweed, and you think of the vast food resources it represents. You have seen great amounts of algae in a fresh-water pond, as well as the vigorous growth of leafy plants floating on the surface or rising above it. There are feasts here for innumerable tiny consumers and for some fairly large ones too.

59

Tropical ferns form a mound
of greenery at the entrance
of Fern Cave in Texas. In caves
green plants survive only
in the twilight zone.

When there is light, there is life

Wherever you direct your eyes—to the mountain slopes and
summits, across the broad prairie, along the tidal salt marsh
or the fresh-water pond, in northern forest or warm savan-
nah—green plants are present. And so long as the sun is shining, these green plants—whether they are unicellular algae,
tiny mosses, slim ferns, durable grasses, flowers, shrubs, or
trees—are engaged in the same business. They are incorpo-
rating the sun's energy into their own expanding substance
and are thereby making food for themselves and for animals.

Green plants are the only organisms that can store the
sun's energy in food. This process is called *photosynthesis*;
the word comes from two Greek words that mean "putting
together with light." All that a plant needs for photosynthesis
is carbon dioxide from the air, water from the soil, and light.

By means of extremely complex chemical reactions carried
on in the parts that contain the green coloring matter *chlorophyll*, plants produce glucose, a simple sugar. This is the
basic food of life. Subsequent chemical reactions, most of

which animals can duplicate, change this sugar into other carbohydrates and into proteins and fats.

Consumers depend on producers

Some animals get their food directly from green plants, and others obtain it indirectly from other animals. *Herbivores*, the plant eaters, get it firsthand by grazing, browsing, gnawing, or chewing parts of plants. The *carnivores* are mainly flesh eaters, consuming creatures ranging in size from microscopic animals to earthworms, insects, rodents, and grazing animals. The *omnivores*, including raccoons, bears, and opossums, enjoy a menu that includes both plants and animals. But all animals are ultimately dependent on green plants for chemical energy and growth materials. They are all *consumers*.

Ecologists, scientists who study the interrelationships between living things and their environments, refer to the passage of energy and materials from producer through a succession of consumers as a *food chain*. The plant that made

Only nongreen plants such as bacteria and fungi grow in the dark interiors of caves; they depend on decay, not photosynthesis, for food. This living mold drapery (a kind of fungus) receives its nourishment second-hand from a disintegrating wooden stairway.

A fungus *(top)* absorbs food from organic debris through thousands of minute threads, or *mycelia*. A dead cricket *(bottom)* nourishes the shroudlike mold encasing its body.

the food is at the base of the chain. The caterpillar that nibbles its leaves forms the first link. The catbird that eats the caterpillar and the hawk that eats the catbird constitute further links.

Decomposers—fungi and bacteria—are the last link. They close the chain, finding their food source in dead plants and animals. The decomposers reduce the dead material to simple chemical compounds that replenish the soil and give back to the air the carbon dioxide originally used by the plants in photosynthesis.

Scientists have studied the aboveground food chain as it operates in a populous, food-rich lake. Copepods, small water-dwelling animals, feed on the algae in the water. In turn, the copepods are eaten by phantom midge larvae, and phantom midge larvae are eaten by dragonfly larvae. Afterward dragonfly larvae are eaten by minnows, which are then eaten by large bass. Thus there are five successive kinds of feeders—five *trophic levels*—from the green producers of food to the fish consumers. Five levels are probably the maximum for any community.

In this lake community, as in nearly every other natural community, at least 75 per cent of the energy at each level is lost. Less than 25 per cent is passed on to the next higher link in the chain. The bulk of the energy used at each level goes into growth and movement; much less goes into organic or body matter that can be eaten by organisms in the next trophic level. Hence the total weight of living matter—the *biomass*—of the copepods in this lake can never be more than 25 per cent of the weight of the algae; and the biomass of the midge larvae can be only 25 per cent of the weight of copepods consumed. This food chain is topped by bass, but the largest bass population that this lake can support is only one-thousandth of the total weight of the algae.

The relationship of producers and consumers in a natural community is sometimes described as a *food pyramid*. It is a fitting description, since it shows that a large energy base is required in any natural community. Green plants, the producers, provide the broad base by manufacturing food for

This cave spider is a predator and lives by eating the animals it catches in its web. The yellow mold growing on the raccoon droppings is a decomposer; by digesting organic waste, it puts back into circulation the nutrients and carbon dioxide stored in its food.

62

the other members of the community. Anything that reduces the output of the producers—drought, fire, or disease, for example—automatically reduces the number of consumers, perhaps to the point that the few big animals at the top of the pyramid can no longer capture enough prey to meet their own daily needs.

Food must be imported into caves

Even though they seem cut off from the rest of the natural world, the creatures that live in a cave's perpetual darkness are just as much subject to this natural design as the caterpillar, the catbird, or the hawk.

But can there be a food chain or food pyramid when the first link or step doesn't even exist? There are no green plants in the darkness of the cave. Animals that live here must get their food in a more roundabout fashion than consumers outside. Actually it is the outside world that must ultimately supply the cave dweller. (A few bacteria found in caves can manufacture food in the absence of light, but they probably contribute very little to the food supply.)

Aboveground the season of greatest consumption is sum-

Trapped after falling into a cave, this box turtle probably starved to death. Even so, it contributed to the food supply of the cave, for scavengers and decomposers have stripped its bones clean.

Animals at each level of a food pyramid feed on animals and plants at lower levels. The number and total weight of organisms at each successive level tend to decrease because living things convert only a fraction of their food into body tissue. The total volume of life in the lake community (represented here by the large pyramid) is much greater than that in the cave community (the small pyramid) because the lake's energy base—green plants—is constantly producing new food during the warm months. The cave's food supply of organic debris, on the other hand, is severely limited and is replenished only once or twice a year.

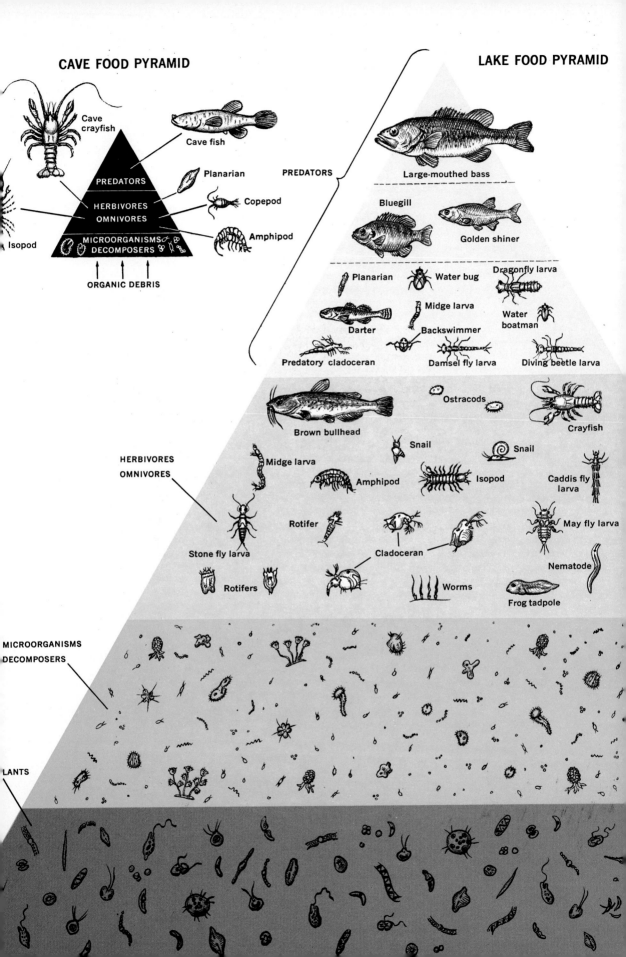

CAVE FOOD PYRAMID

Cave crayfish

Cave fish

Planarian

PREDATORS

HERBIVORES
OMNIVORES

Copepod

Amphipod

Isopod

MICROORGANISMS
DECOMPOSERS

ORGANIC DEBRIS

LAKE FOOD PYRAMID

PREDATORS

Large-mouthed bass

Bluegill

Golden shiner

Planarian   Water bug   Dragonfly larva

Midge larva   Water boatman

Darter   Backswimmer

Predatory cladoceran   Damsel fly larva   Diving beetle larva

Brown bullhead   Ostracods   Crayfish

Snail   Snail

HERBIVORES
OMNIVORES

Midge larva   Amphipod   Isopod   Caddis fly larva

Rotifer   May fly larva

Stone fly larva   Cladoceran   Nematode

Rotifers   Worms   Frog tadpole

MICROORGANISMS
DECOMPOSERS

LANTS

FOOD WEBS AND SPECIALIZATION

The arrows in these food webs represent the flow of food through two communities. The width of the arrows indicates the relative importance of each food source in an animal's diet. By comparing the food relationships of one animal from a cave community and one from a lake community, you can easily see how the relative abundance or scarcity of food has affected their eating habits. In a food-rich lake, for example, the golden shiner depends on several food sources, yet large copepods are by far its most important prey: the shiner is a food specialist. The cave crayfish, on the other hand, depends on many food sources, all more or less equal in importance. Like most cave animals, it is omnivorous, eating whatever food it can find. It cannot afford to be a specialist.

ENERGY OF THE SUN

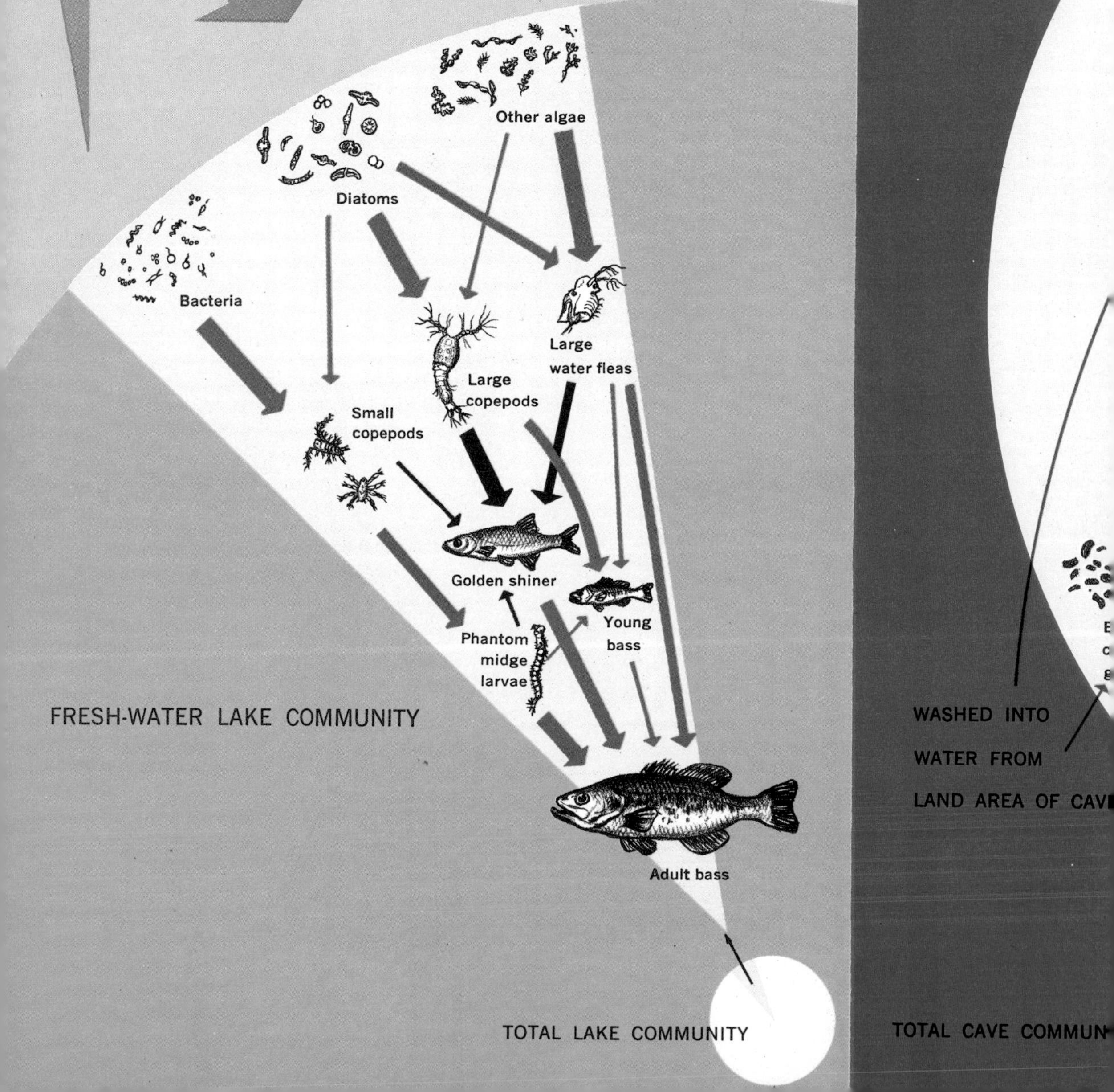

Other algae

Diatoms

Bacteria

Large water fleas

Large copepods

Small copepods

Golden shiner

Young bass

Phantom midge larvae

FRESH-WATER LAKE COMMUNITY

Adult bass

TOTAL LAKE COMMUNITY

WASHED INTO WATER FROM LAND AREA OF CAVE

TOTAL CAVE COMMUNITY

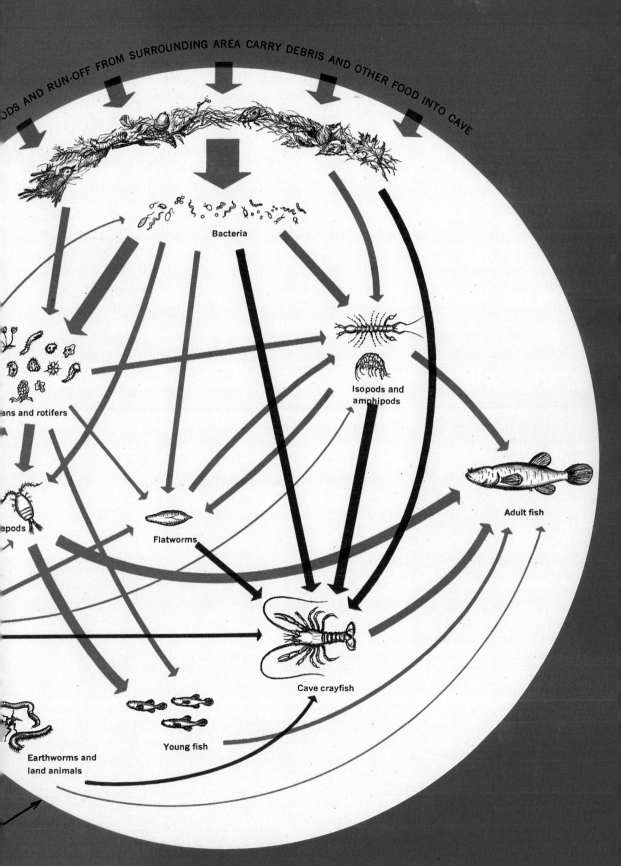

FLOODS AND RUN-OFF FROM SURROUNDING AREA CARRY DEBRIS AND OTHER FOOD INTO CAVE

Bacteria

...ans and rotifers

Isopods and amphipods

...epods

Flatworms

Adult fish

Cave crayfish

Young fish

Earthworms and land animals

mertime, when animal populations are increasing and plants are at peak production, turning out leaves, fruit, and seeds. Even the roots and stems store energy. But the great quantity of organic material that accumulates aboveground in autumn does not benefit cave dwellers unless some of it is carried into the caves.

Organic material may be brought into caves by animals, either accidentally or on purpose, or it may be carried in by floods. If an animal falls into a sinkhole and dies, it is potential food. If it wanders in, as raccoons often do, and leaves droppings, these are also potential food. Because they regularly roost in caves, bats may leave thick carpets of droppings. However, even a large accumulation of their droppings represents a minute quantity of food compared with that available in the outside world. Floods, too, bring in relatively little food.

Even the most food-rich caves, then, are poorly supplied compared with aboveground communities, and so it really is not surprising that caves are sparsely populated, that they support only a handful of species. What little food is imported into caves must sustain all vital activity there: the birth and growth of each individual; the movements of legs, feelers, and jaws; the breathing and heartbeats; the pursuits, captures, and escapes.

No gourmet, a blind isopod makes the most of the meager pickings in a cave. Dead organic material, bacteria, protozoans, flatworms, and even guano contribute to its diet.

Cave animals are seldom specialists

In food-rich areas aboveground, animals are likely to be specialists, eating only certain kinds of plants or other animals that are abundant and easy to find. But most cave dwellers cannot afford to be "choosy." They cannot by-pass nearby food and squander energy in pursuit of possible meals farther away. Cave animals, in short, must be opportunists and take advantage of whatever food is available.

Where food in a cave is relatively plentiful, there are, at best, three food levels. But nearly all the cave's inhabitants are omnivorous. The same species of animal may limit itself to one kind of food outside of the cave but eat several kinds inside. Outside, for example, the snail *Oxychilus* eats nothing but plants. But inside the cave this snail is more adaptable: it adds animal carcasses as well as bat droppings to its menu.

This latter substance—bat *guano*, as it is called—ranks with debris as a major food for cave animals. In many communities, at certain times of the year, the build-up of guano under a roost can support a large population of cave animals.

Bats bring food into caves

Since northern winters clear the skies of insects, most species of bats exist on a starvation diet of their own fat. The amount of guano deposited during their long winter of hibernation is hardly noticeable on a cave's floor. In summer most species of bats desert their winter roosts in caves and scatter over the countryside.

A few species of bats, however, do colonize caves during summer in certain regions. In Alabama, Georgia, Missouri, Arkansas, Oklahoma, and states farther west, the "gray bat," *Myotis grisescens*, occupies caves in summer as well as in winter. Mountains of guano pile up beneath the bat roosts, and each cave is known locally as Bat Cave.

A characteristic odor, resulting from the accumulation of droppings and urine beneath the roots, often reveals the presence of such a cave. That is how author Charles Mohr found Bat Cave in the Ozark Mountains of northwestern Arkansas. Descriptions of the route to the cave were so vague that he searched along the limestone cliffs for an hour before an unmistakable odor informed him that the cave was nearby. Pushing through a screen of underbrush, he found the entrance.

He had not moved very far down the main passage when his path was blocked by a lake extending from wall to wall. The air was full of bats flying from clusters hanging over the water. Within minutes the whole colony took off and disappeared down a dark corridor beyond the lake. The dull echoes of their wingbeats could still be heard after they were out of sight.

Mohr quickly forgot the bats because as he reached the edge of the lake, he found himself sinking into a thick, reeking brown substance, not as solid as mud nor as liquid as soup. The bottom of the shallow lake and its shore were one huge guano deposit, ages in the making.

Swimming on the surface of the water and crawling over the guano on the bottom were thousands of white worms, slender and paper-thin. They were flatworms, or *planarians*, creatures that are particularly interesting to biologists because of their truly remarkable powers of regeneration. If

Unusually abundant, these flatworms in a pool in an Oklahoma bat cave may feed directly on guano or, just as likely, prey on smaller animals attracted by the rich food supply.

69

At sunset half a million free-tailed bats emerge from their daytime roost in a remote section of Carlsbad Caverns. These bats fly half a mile underground and then as far as fifty miles from home on their nightly insect hunts.

they lose an organ, they can grow a replacement. If the head is cut off, it grows a tail, and the tail of the original flatworm grows a head.

Aboveground, pigmented planarians with eyes are found commonly in springs and streams, where they generally hide under stones. The ones in this cave were swimming at or near the water surface, and they were white and eyeless. In other caves, planarians are encountered only rarely; seldom are as many as a dozen seen. In this pool there were tens of thousands. Those that were thin as fine thread were the young ones. Many adults were as much as half an inch long.

The planarians on the surface were swimming in the manner of this species, *Sphalloplana percaeca*, gliding on the water's surface film. Most were submerged and were crawling over dead crickets, isopods, and pieces of wood. Mohr could not be certain whether or not they were taking their nourishment directly from the guano, but clearly they were profiting, if only indirectly, from its presence.

Later, in Tumbling Rock Cave, in northeastern Alabama, Mohr found land animals living directly on bat guano. In addition to pale brown blind beetles, he saw great numbers of earthworms, snails, and millipedes. They were swarming over the freshest guano, which had a sweetish, less repugnant odor than the liquid guano in Bat Cave and resembled partly cooked rice in its shape and consistency. Despite its origin and appearance, guano has a high nitrogen content and is an excellent food for animals of many kinds.

The guano caves of the Southwest

The biggest guano deposits reported by speleologists have been in certain caves of Texas and New Mexico in which the "free-tailed" or "guano bats," *Tadarida brasiliensis*, spend the summer months. Most of the truly tremendous bat colonies—numbering as many as 30 million—are in remote, nearly inaccessible places.

A major exception is Carlsbad Caverns in Carlsbad Caverns National Park in New Mexico. In summer visitors stay until dusk to watch as many as half a million bats rush from the huge main entrance and disappear in the direction of the Black River, four miles away. The public sees the bats only during their flights from the cave. Their roosting area is half

The free-tailed bat, so-called because part of its tail extends beyond the edge of the tail membrane, is the country's most abundant bat species. Nonhibernating, it migrates south in winter and returns each summer to congregate by the millions in caves throughout the Southwest.

CARLSBAD CAVERNS NATIONAL PARK

The world-famous caves of Carlsbad Caverns National Park, in southeastern New Mexico, contain the largest and most magnificent underground chambers known to man. The largest one of all, called simply the "Big Room," is 1800 feet long, with a floor space of fourteen acres and a ceiling that arches 255 feet overhead. Like all the other rooms and passages in the cave system, it is studded with elaborate, delicately tinted formations. Stalactites and stalagmites seem to sprout from every inch of floor and ceiling space. Flowstone, drapery, grape formations, and helictites mingle everywhere, creating scenes of unequaled beauty. The Giant Dome is one of a group of huge stalagmites that tower above the tourist trail *(left)*. The massive Christmas Tree *(right)* has been seen by relatively few people since it is in New Cave, which is located about twenty miles from Carlsbad Caverns. New Cave is within the park but is not open to the general public. On the next two pages is the Caverns' immense natural entrance, used by a colony of half a million free-tailed bats.

a mile off the tourist route, reached by climbing and crawling along a seldom-traveled route.

A trip into any of the Southwest's bat caves is not for the squeamish. In summer the heat is stifling, and the ammonia fumes are so strong that the fur of some bats is bleached a pale brown. The roar of a hundred thousand beating wings is almost deafening as part of the colony takes flight.

Ceiling area in the larger bat caves can be measured by the thousands of square feet, yet scarcely an unoccupied space can be seen. There is a continuous rain of urine and feces and even dead and dying bats, most of them too young to fly. Still more disturbing are a fine, steady rain of mites from the roosting bats and the swarms of blind flies that live as bat parasites. The surface of the vast carpet of guano

76

stirs with life, chiefly beetles and moth larvae. Wherever you look, insects are feeding on the carcasses of young bats. Thousands of bat skeletons, picked clean to the last shred of flesh, lie all about. Moth larvae, of a species related to the clothes moth, are eating the hair shed by the bats during their periodic moltings.

When Mohr visited Carlsbad Caverns, the park naturalist was studying the remarkable moth population in one portion of the cave, and he pointed out the webbing spun by the larval moths among the bat hairs and guano. This guano bed had accumulated in a mere twenty years, he said. Prior to 1941, when Carlsbad Caverns National Park was established, miners had taken out more than 100,000 tons of these nitrogen-rich deposits of guano and sold it for fertilizer.

So plentiful that their colonies often number in the millions, free-tailed bats may cluster in a single "rug" spreading across thousands of square feet of cave ceiling. Each of these half-ounce bats consumes nearly half its weight in insects in a single night. Thus the guano that accumulates beneath a roost in the course of a summer may provide hundreds of tons of food for myriads of cave dwellers.

Bats are attacked by an amazing
assortment of ticks, mites, fleas, and
many other parasites, such as the tiny
bloodsucking fly shown clinging to a
bat's wing membrane *(above)*. But since
bats are fastidious and constantly groom
their fur, they seldom harbor more
than a few parasites at a time.

In addition to the droppings, urine, and
evicted parasites constantly showering from
a huge bat roost, the bodies of weak or aging
bats and occasional newborn young add to
the banquet spread for scavengers on the cave
floor. Beetle larvae and a host of other
animals soon pick the bodies clean, leaving
the floor littered with whitened skulls
and slender wing bones.

78

Confederate saltpeter caves

Many caves that once harbored tremendous bat colonies are now deserted. The bats probably left the caves as the climate became increasingly dry and the insect populations on which they fed diminished.

In the deep South, dry caves were a major wartime resource to the Confederacy. Miners excavated cave earth and poured water through it to recover the calcium nitrate—a process known as leaching. By boiling the calcium nitrate solution with potash, they obtained the saltpeter (potassium nitrate) they needed for making explosives. Gunpowder acquired in this way accounted for half the munitions used by the Confederates, who otherwise might have been forced to surrender years earlier, so effective was the Union blockade of southern ports.

Visitors to Sauta Cave, near Scottsboro, Alabama, have little trouble visualizing how the cave looked in the days when gunfire was booming across the Shenandoah Valley and Sherman was marching to the sea. The leaching vats and scaffolding used by the saltpeter miners are still remarkably intact. Small metal railroad cars sit astride tough wooden rails which are so well preserved that they could still serve the cause that was lost a hundred years ago. The molds and fungi that by now would have rotted the rails beyond recognition probably were killed by the dryness that may have been responsible for the bats' departure. And now there is no community of animals because food is no longer brought into the cave.

Crickets bring food into caves

To study all the links in the cave's interconnecting food chains, scientists must make repeated trips into the caves and spend hundreds of hours surveying, observing, and analyzing the animals and their habitat. In the past, scientists have concentrated on separate studies: the origin of caves, their physical arrangement, the habits of bats or fish, and so forth. Finally, in 1960, a comprehensive study of a total population of animals in a small but richly populated Kentucky cave was begun by Brother G. Nicholas, a former president of the National Speleological Society.

He chose Cathedral Cave in Kentucky, high up on a bluff

During the Civil War mule-drawn carts carrying nitrous earth rumbled down these wooden tracks in a southern cave. Saltpeter refined from the earth was used to make gunpowder for the Confederate armies.

overlooking the Green River in Mammoth Cave National Park. There are no blindfish or crayfish there, but almost every other kind of cave animal has been observed: isopods, amphipods, flatworms, earthworms, crickets, beetles, land snails, spiders, harvestmen, millipedes, and others.

Brother Nicholas or one of his three college-age assistants visited the cave daily for three years. They started by marking off the cave into twelve sections, each ten feet in length. Whenever an insect or other invertebrate was found, it was dabbed with colored paint, and its position was noted. Twelve different colors were selected to represent the twelve sections. On each of the 1100 days of the experiment, an observer carefully noted the position of all the animals, marking any new ones not already painted. Several species of cave animals were present, but the overwhelming majority of individuals were crickets. There were 3750 common cave crickets (*Hadenoecus subterraneus*) and about 700 permanent cave dwellers, mostly beetles.

Brother Nicholas and his assistants accumulated a staggering mass of information on the activity of the color-marked insects. Their observations showed that at dusk every night, except when the temperature was below freezing or the relative humidity had fallen below 85 per cent, one-third of the *Hadenoecus* colony emerged from the cave, presumably on a feeding expedition. Before dawn the foragers returned with full guts to their proper places in the cave. This third of the population remained home during the next two nights, while the rest took their turns to go outside.

In the period when the insects were under observation, close to a quarter of a million individual foraging trips created a fantastic volume of traffic. The Nicholas investigation proved that *Hadenoecus* was the cave's only important source of food from outside. Cathedral Cave has no bat population and no large sinkholes to trap significant quantities of organic debris. While the cricket guano in Cathedral Cave may build up to depths of half an inch, it is insignificant compared with the guano left by active bat colonies, and so it is not surprising that there were only two major trophic levels in the cave: the 3750 foraging crickets and the much less numerous permanent cave dwellers that lived directly or indirectly on cricket guano. In terms of biomass there were about forty times as many crickets as permanent cave dwellers.

A given amount of guano has much less food value than

By daubing individuals of a cricket colony with fast-drying paint, biologists were able to trace their movements in one of Mammoth Cave National Park's smaller caves. Their three-year study revealed that about one-third of the colony went outside each night to feed.

Large scorpions, such as the three-inch specimen shown here, often lurk in Texas caves. Adept at hunting live crickets, the agile predators first sting their victims with the poison-injecting barbs at the tips of their tails and then suck out the body juices.

Quick to locate dead crickets, half-inch-long rove beetles seem to come from nowhere and flock to the feast. Even so, they occasionally arrive too late; live crickets sometimes beat the beetles to the food.

LIFE IN A CRICKET CAVE

By making almost nightly trips to the surface, where they gorge on plant food, cave crickets frequently support a surprising variety of camp followers. In a Kentucky cave, for example, Brother G. Nicholas found relatively large populations of permanent cave dwellers living on cricket guano, which occasionally accumulated in layers as much as half an inch thick. Robert W. Mitchell of Texas Technological College confirms the importance of *Ceuthophilus* crickets (actually a kind of grasshopper) in some southwestern cave communities. In the semiarid Edwards Plateau in central Texas, however, cricket guano is a relatively unimportant food source, but the crickets themselves are eaten by many other members of the cave community. Large scorpions lurk in dim passageways, ever alert for passing crickets. Tiny rhadinid beetles scout the floor in an endless search for cricket eggs. When a cricket loses a jumping leg—a fairly common occurrence—harvestmen are quick to discover the prize. Dead crickets are rapidly consumed by rove beetles, springtails, harvestmen, and even other crickets. If scavengers should overlook a dead body, a fuzzy growth of mold soon covers it, and other animals feed on the mold. And so it goes. In every stage of life and death, the cave crickets are at the base of the cave food pyramid; without them, most of the other inhabitants of the cricket cave would quickly starve.

On the next two pages, a harvestman clinging to the underhang of a large rock tackles a meal almost as big as itself—a cricket's jumping leg, which it probably discovered on the cave floor. Two red chiggers, tiny larvae of a parasitic mite, cling to the harvestman's body and one of its legs.

1

Tapping with its antennae and
probing with its shorter maxillary
palps, a blind rhadinid beetle
prospects for buried treasure (1).
Its sense of touch and the chemical
receptors on its antennae and
palps lead it to hidden cricket
eggs. Using its head like a hoe,
the beetle scrapes a hole in the
silt and uncovers an egg (2).
Clamping the egg in tonglike
mandibles, it punctures the tough
shell (3) and pumps the egg's
nutritious contents into its
mouth (4). Like a crumpling bag,
the membrane slowly collapses.
Finally, the beetle becomes bloated
from eating an egg nearly as big
as itself (5).

2

BEETLES THAT EAT CRICKET EGGS

Curious about the origin of tiny holes, like
pockmarks, on the silt-covered floors of certain
Texas caves, Robert W. Mitchell set out to
investigate. Two years of observation and
experiment eventually confirmed his suspicions.
The holes are the handiwork of the beetle
Rhadine subterranea. Surprisingly specialized
for an animal that lives in caves, the blind
quarter-inch-long beetle feeds almost exclusively
on cricket eggs. Every detail of its behavior
and anatomy seems designed for the work of
locating and stealing eggs that the crickets
deposit one at a time beneath the surface of
the silt. Sensory organs on its antennae and
maxillary palps—club-shaped appendages on
each side of its mouth—are the biological
"Geiger counters" that lead the beetle to its food.
Chemical substances left by the cricket
reveal the presence of recently laid eggs.
Minute irregularities on the surface of the silt
may be the clues that lead the beetle to older
eggs. Once it has located the spot where an
egg is buried, the beetle digs a hole, pulls out
the egg, and then devours its contents.

5

the same amount of live animal. If the live crickets in Cathedral Cave were at the base of the food pyramid, instead of their guano, perhaps there could be more permanent cave dwellers. In Mammoth Cave, Thomas Barr discovered huge populations of blind cave beetles—*Neaphenops*—living with crickets in remote upper passages. Observation showed that the beetles were eating cricket eggs. Here the biomass of the crickets was only about ten times that of the beetles. By eating cricket eggs, *Neaphenops* is able to maintain much larger relative populations than other beetles which prey on cave animals that get their food second-hand from the cricket guano or from debris washed in by spring floods.

Timing of reproduction

Unfortunately for the *Neaphenops* beetles, cricket eggs are not available all year long. Crickets lay their eggs when they can feed outside, chiefly in summer and early fall. Only then do they get enough surplus food to produce eggs. And similarly the beetles cannot lay eggs unless there are cricket eggs to eat. Thus, like the crickets, *Neaphenops* must be a seasonal breeder. But how do the crickets know what season it is? After all, caves have a constant climate compared with the world outside. What are the clues telling the crickets that spring has come and they can start their nightly foraging trips again?

All cave insects are very sensitive to changes in moisture; their body walls are thin, and they lose body water more easily than most noncave species. Thus they probably can sense the seasonal variations in the rate at which water evaporates in caves, and perhaps it is just such subtle changes as these that synchronize their behavior with the progress of the seasons outside.

But of course, some seasonal changes are not so subtle. In regions where the ground is frozen or is saturated by winter rains, the normal snowmelt and spring rains sometimes produce severe flooding. If steep terrains cause rapid run-off

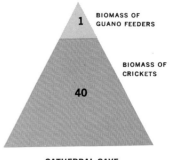

BIOMASS OF GUANO FEEDERS

1

BIOMASS OF CRICKETS

40

CATHEDRAL CAVE

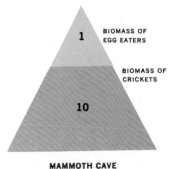

BIOMASS OF EGG EATERS

1

BIOMASS OF CRICKETS

10

MAMMOTH CAVE

The eggs of crickets are far richer in food value than their guano. As a result, a population of cave crickets can support a greater biomass, or volume, of egg-eating beetles than of animals that feed only on the less nutritious guano. In Cathedral Cave, for example, the biomass of guano-eating beetles and other permanent cave dwellers is only one-fortieth of the biomass of the crickets. On the other hand, the biomass of egg-eating beetles inhabiting the upper passages of Mammoth Cave is about one-tenth the biomass of the crickets.

Seldom numerous enough to support a retinue of guano feeders, the cricket *Ceuthophilus* lives nearer cave entrances than *Hadenoecus*. Larger eyes, prominent markings, and stouter body readily distinguish it from its pale, nearly blind relative.

into caves, rushing floodwaters carry in all sorts of debris—seeds, leaves, twigs, bits of bark and broken branches, occasional logs, and the remains of drowned animals, from ants and worms to the carcasses of lizards and mice.

The debris carried into a cave by spring floods affects aquatic animals in the following way: Bacteria and protozoans feed on the organic matter suspended in the water and deposited on stream bottoms, and their populations build up very rapidly. Next, copepods, isopods, and amphipods hatch just in time to take advantage of the build-up of bacteria and protozoans. Finally, crayfish and fish hatch in midsummer from eggs laid when the spring floodwaters were at their highest levels; these young animals are on their own just in time to feed on the peak populations of young copepods, isopods, and amphipods.

It is one thing, however, merely to recognize that cave fish and cave crayfish lay their eggs at the time of the spring floods; it is quite another thing to explain why. The fact that it is advantageous for them to do so is not a good enough answer. The scientist wants to know what "triggers," or starts, the series of changes that occur in the animals'

bodies prior to the actual egg laying. These changes take place *before* any sign of flooding is evident—when the cave environment is in its most stable condition. One reasonable explanation for the changes is that the fish and crayfish, like all creatures living aboveground, possess internal time-measuring mechanisms, or "biological clocks."

The crayfish in Shiloh Cave, Indiana, mature and develop eggs and sperm at the same time every fall, long before the spring floods. Thus the females are prepared to lay eggs whenever the spring floods come. Each year the crayfish lay their eggs at a different time, exactly coinciding with the time of maximum flooding. Apparently their "clocks" control the production of eggs and sperm, and the floods trigger egg laying.

It is remarkable that cave crayfish, and cave fish as well, have kept their biological clocks going. There are no twenty-four-hour clues in the cave. The permanent cave dwellers do not make regular expeditions to the outside world, so their clocks cannot be reset periodically in the same way as those of cave crickets or bats. Yet there is good reason for the cave animals not to have lost their biological

Spring floods usually are the most important event of the year in many caves. Besides replenishing the supply of organic debris, floods seem to be the signals that trigger the egg laying of many cave dwellers.

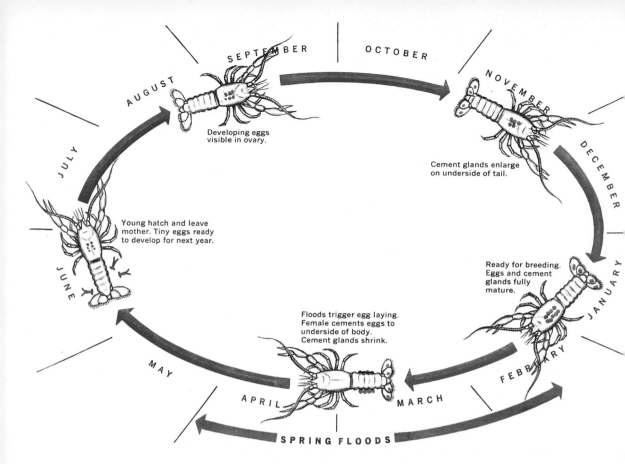

SEPTEMBER · OCTOBER · NOVEMBER · DECEMBER · JANUARY · FEBRUARY · MARCH · APRIL · MAY · JUNE · JULY · AUGUST

Developing eggs visible in ovary.

Cement glands enlarge on underside of tail.

Ready for breeding. Eggs and cement glands fully mature.

Floods trigger egg laying. Female cements eggs to underside of body. Cement glands shrink.

Young hatch and leave mother. Tiny eggs ready to develop for next year.

SPRING FLOODS

Internal time-measuring mechanisms, or "biological clocks," prepare crayfish for the breeding season. Without any clues to time of year, females in Shiloh Cave, Indiana, invariably are ready to lay their eggs during spring flooding—even though the time of flooding varies from year to year. Males go through a similar cycle of preparation before each breeding season.

clocks. The clocks allow them to make best use of what little food comes into the cave every year. And the amount of food that washes in one year determines how many animals will breed the next year.

How cave fish populations are regulated

Our knowledge of how cave fish control their numbers is based partly on a seven-year study of what might properly be called "Blindfish Cave," in Indiana. Actually, it is identified on maps as Upper Twin Cave and is located in Spring Mill State Park, about fifty miles south of Bloomington.

The investigation, by author Thomas Poulson, centered on the largest cave dweller, the four- to five-inch-long cave fish *Amblyopsis spelaea*. As the largest carnivorous species in the cave, it is at the top of the food pyramid, where the effects of any population changes among animals at lower levels of the pyramid can be most easily detected.

Fortunately, the cave and its inhabitants are rigorously guarded by the Indiana State Divisions of State Parks and

Fish and Game. Both for the safety of the cave animals and for the validity and significance of the scientific findings, complete protection is essential.

Upper Twin Cave is about 4500 feet long and can be explored only by those willing to swim in 56° water and climb over slippery rock ledges. It takes ten to twelve hours to count all the fish. One 250-foot section has an unusually large number of fish compared with other areas in the cave. About 20 of the 120 fish making up the population are found here.

The biggest fish are in the deeper water, especially along underwater ledges. Each swims and feeds regularly in a limited area, moving upstream along a ledge for about sixty feet, swimming to the surface where the current is swifter, then floating downstream to its starting point, and finally starting upstream along the ledge again. As a fish moves upstream along the ledge, it noses under rocks. Poulson learned through careful observation that the time a fish spends under a rock is directly related to the number of isopods to be found there.

In early winter, when the water generally is still clear, the larger fish appear to be spread out along the stream passage in pairs, male and female. Each pair seems to stay in its own area, but we do not know yet whether the fish defend their territories, as birds do.

A biologist matches wits with blind cave fish. Although easy to approach, the white three- to four-inch-long fish are difficult to capture. They are amazingly sensitive to vibrations in water and often manage to elude the approaching net.

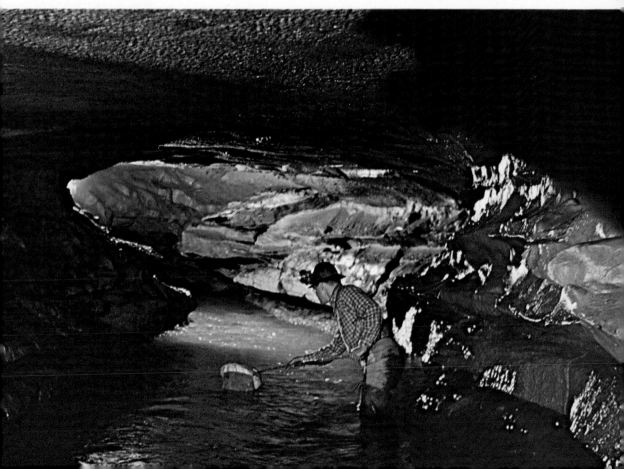

During the seven-year study of Upper Twin Cave, there were never less than 81 fish, nor more than 130. Apparently the population is very stable; there are no wasteful population build-ups, no excessive losses. But how is this stability achieved?

As the population reaches its peak, there is such competition for food that the adult fish get barely enough to stay alive. Very little extra food is available for eggs. (To produce eggs, females need three times as much food as usual.) On the other hand, when there are fewer adults and less of the available food is eaten, the females can find enough food to produce eggs. This food enables some females—on the average, about five individuals a year—each to produce forty to sixty large, heavily yolked eggs. The other females, who are

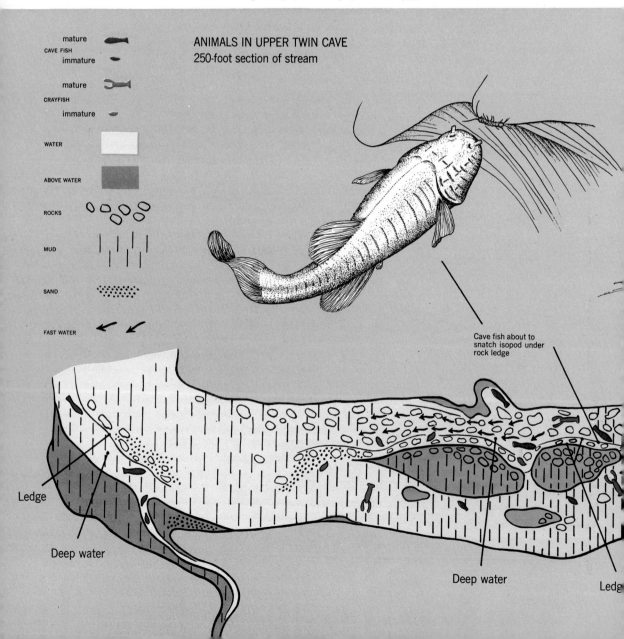

ANIMALS IN UPPER TWIN CAVE
250-foot section of stream

mature
CAVE FISH
immature

mature
CRAYFISH
immature

WATER

ABOVE WATER

ROCKS

MUD

SAND

FAST WATER

Cave fish about to
snatch isopod under
rock ledge

Ledge

Deep water

Deep water

Ledge

not so fortunate in obtaining surplus food, do not breed at all.

If a year of better-than-normal food supply coincides with a year of low adult population, more females than usual produce eggs. The increased number of young may cause trouble: they compete with adults for food. A year later, if the food supply drops back to normal, or below, there is a natural way of reducing the number of young fish: cannibalism! Of course, the larger fish do not *consciously* set out to reduce the population; they are cannibals merely because the young fish move around much like normal prey. However, the very small young fish are not completely without defense; they react to any disturbance in the water by "freezing" and do not move even when touched.

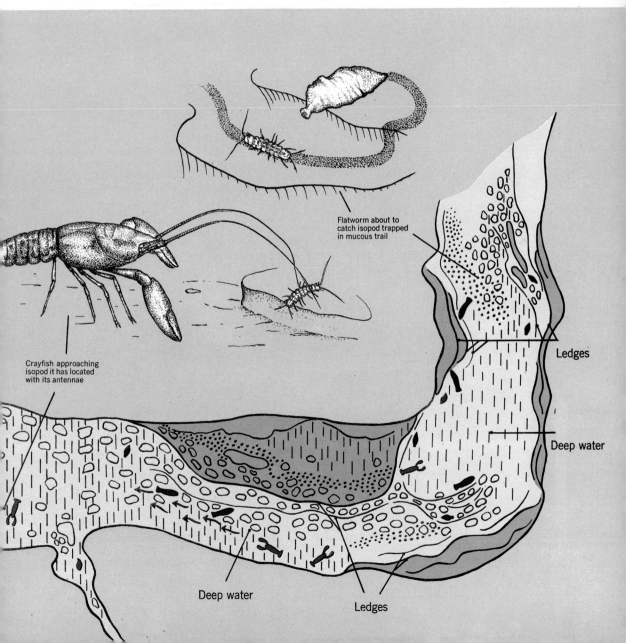

Flatworm about to catch isopod trapped in mucous trail

Crayfish approaching isopod it has located with its antennae

Ledges

Deep water

Deep water

Ledges

Food: the limiting factor in caves

In every natural community the sun provides the energy that green plants, the producers, need to manufacture food. Animals, the consumers, get their energy second-, third-, or fourth-hand from the plant producers. Cave animals are no exception to this rule.

Since there are no green plant producers in caves, permanent cave dwellers must depend upon food brought in from the outside by animals or by floods. Compared with aboveground communities, no caves have an abundance of food, and the supply is likely to be replenished only once a year. Cave animals, therefore, are sparing and efficient in their use of energy and hence are able to endure long periods of deprivation. Their reproduction is keyed to the times of plenty, and their population is so closely regulated that it does not outstrip the meager food supplies. In meeting and overcoming these harsh realities, cave dwellers have developed truly remarkable adaptations, as the next chapter shows.

Seemingly meager fare, a tangle of rotting twigs, seeds, bits of bark, and other flood-deposited debris are a bonanza for a blind millipede. Like all dwellers in the perpetual darkness of the cave, it depends for its very existence on food carried into the cave from the sunlit world aboveground.

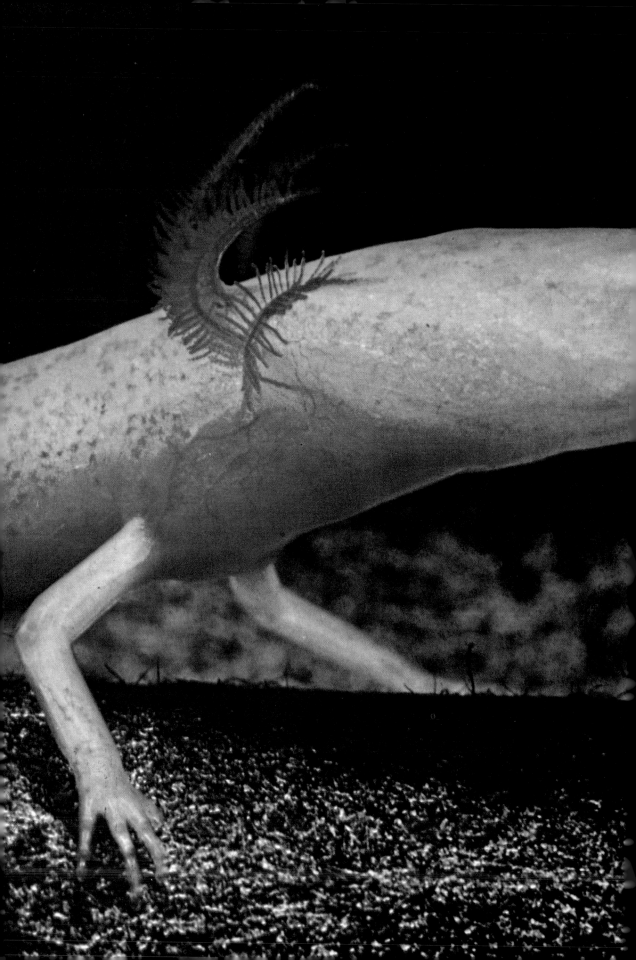

Life in

Darkness

What is so remarkable about cave animals living in the dark?
If you have been outdoors much at night, you must have
seen, heard, or otherwise sensed the presence of a host of
nocturnal animals. Prowling the forest, creeping through the
brush, darting across the meadow, patrolling the water's
edge, or hovering high above the ground, they seem unaffected by the absence of daylight.

Naturalists tell us that a nighttime census of a swamp, a
forest, or a desert may yield more species and more individuals than could be found by day. Certainly much insect
activity occurs at night, as any camper will testify.

Night is the best feeding time for many mammals. Rab-
bits, flying squirrels, and mice come out of their hiding
places and forage. A doe with its fawn steps softly through
the dark, nibbling green twigs at the edge of the forest. The
opossum clambers down from his daytime perch to hunt for
blackberries and other wild fruits.

All these animals depend to some degree on sight. The
light of the moon, or even the diffused starlight, is enough to
guide them through the forest, enabling them to avoid obstacles and to find their food. Some have eyes hundreds of

99

Common in parts of the Southwest, the cat-sized ringtail is so shy that it is seldom seen. Emerging after dark from its den in a hollow tree, rock crevice, or cave, it slinks silently through the night in search of mice, rats, insects, and almost anything else that is edible.

times more sensitive than our own. But it is other sharp senses—hearing, smell, even touch—that come into full play at night.

The damp, motionless air holds odors close to the ground, and food can be smelled from a distance. Sounds carry farther and more distinctly than they do during the day. The rustle of grass, the whir of wings, or the snap of a twig is instantly perceived as a signal for alertness or flight.

Many predators also are abroad in the darkness. Foxes pad noiselessly through the clearings. Lynxes and wildcats crouch on the limbs of trees, ready to pounce upon small forest creatures. Bats dart and veer in pursuit of moths and other flying insects. Whippoorwills, camouflaged during the day, now leave their hiding places. Hungry owls are poised and waiting. Tree toads and salamanders, which must spend the day in damp places to keep from drying out, can move safely through the moist night, preying on insects. Raccoons search the rocky stream bed for crayfish.

The night is alive with the cries, warnings, and invitations of unseen presences. In the breeding season, mockingbirds

100

and song sparrows often utter their bold notes and trills. Whippoorwills and chuck-will's-widows call repetitiously, and loons shriek their haunting laughter. The metered chirps of the snowy tree crickets are tuned by the temperature. A staccato chorus of frogs and toads arises from woodland ponds and marshes.

But the cave certainly is one of the most silent habitats on earth. There is no rustle of moving bodies, no small telltale sounds—nothing but the occasional splash of dripping water and the protesting squeak of a roused bat, or its fluttering wingbeat. There are no birds calling in the darkness, no chirring of insects.

The cave, with its continual darkness and its protective humidity, would seem a natural headquarters for many nocturnal creatures. Yet, except for bats, few warm-blooded nocturnal species make regular use of it. Pack rats often nest there, but they, like the bats, are only part-time guests. Phoebes and, on rare occasions, owls nest in the cave's twilight zone, close to the entrance. But even though they may visit the cave from time to time, most night-faring animals look for food and mate and carry on their vital activities out-

The timid flying squirrel, with eyes so big that it seems forever frightened, seeks nuts and other seeds at night. A furry fold of skin along each side of its body helps this little rodent to glide noiselessly from treetops to forest floor.

side. Perhaps they are not able to adjust themselves to the cave's utter darkness, in which even the keenest eyes are useless.

How does an owl find its prey?

Of all nocturnal warm-blooded creatures, owls would seem to have the best chance to hunt successfully inside a cave. For a long time some naturalists believed that owls could catch their prey in total darkness, but they had no way of proving it. It remained for Roger Payne, a biologist at Cornell University's Laboratory of Ornithology, to adapt military infrared equipment for watching owls as they catch their prey in the absence of any light.

Placing captive barn owls in a huge shed, Payne studied them as they pursued mice he had released. First he studied them under different light conditions, some fairly bright, some quite dim. Over and over again he repeated the experiment, carefully recording the intensity of illumination.

In every case, the owl attacked in the same way. The moment it sighted its quarry, it glided toward it directly on outspread, motionless wings, feet thrust backward against its tail. When it reached a position above the mouse, it would throw back its head, bring its feet forward with talons spread for the seizure, abruptly raise its wings, and drop in a split second on its victim.

Now that he knew exactly how an owl attacks a mouse that it sees, Payne was ready to repeat the experiment in the dark. In a completely lightproof building he released a mouse and freed the owl. Using a "sniperscope," an apparatus projecting an infrared beam to which the owl's eyes are insensitive, and a movie camera shooting infrared film, he was able to study and record how the owl captures prey it cannot see.

In the first of these experiments, Payne released the mouse on a floor littered with leaves and grass to simulate woodland conditions. Pictures taken on the infrared film showed that at the moment the mouse was released, the owl instantly faced about in that direction, just as it did when

Spelunkers sometimes find raccoon tracks in caves, far from any known entrances. Analysis of the scats indicates that these inquisitive mammals occasionally eat cave crayfish, which they probably catch by "fishing" with their paws.

This ten-inch-tall screech owl roosts in the twilight zone of a Missouri cavern. Though skilled in night flying, owls seldom venture far into caves. Even their remarkably sensitive vision is useless in the utter blackness beyond the twilight zone.

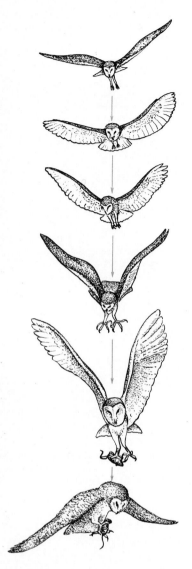

A BARN OWL POUNCING UPON A MOUSE

it could see the mouse. But its flight was markedly different. Instead of gliding on rigid outstretched wings toward its quarry, it flapped its wings continuously as it approached. The swift beat of the owl's wings set its body rocking; its feet swung back and forth like a pendulum. In the final instant, as it closed in for the kill, it thrust its talons forward directly above the mouse. If the mouse swerved, the owl followed in a split second. It heard and followed its victim through every shift. No mouse escaped.

Payne released other mice on a sand-covered floor. Even when the mice were made to squeak at the moment of release, the owl could not find them. The sounds of their movement were so reduced and deadened that the owl could not determine their position with accuracy. Every mouse in this experiment made its escape, and in a short time the owl gave up looking for mice under these conditions.

To test his theory that hearing alone enables an owl to find its prey, Payne made one further experiment. He tied a small wad of paper to a thread and dragged it across a pile of leaves. The owl caught the wad of paper immediately. Since the paper had no animal odor, obviously the owl was not finding it by scent. Since it generated no heat, it was clear that the owl did not rely on heat-sensing organs such as rattlesnakes possess. There could be no doubt that the barn owl's hearing was so acute that it could find and capture its prey without the use of other sense organs.

Would an owl's supersensitive hearing work in a cave? It probably would not, except in entranceways where there might be enough leaves to produce a rustling sound if a mouse moved. Owls apparently feed exclusively outside of caves. However, they may catch an occasional bat as it emerges from a cave during the night; owl pellets containing bat bones have been found in quite a few cave entrances where the birds apparently came back to perch while consuming their prey.

Cave-dwelling birds

A few North American birds occasionally nest in caves. Phoebes and both cave and cliff swallows have been known to fasten their mud nests to vertical rocky entranceways. Barn owls and several species of wrens sometimes build their nests in small pockets or crevices in the rock. But

104

these random guests are never found beyond the twilight zone. Their nesting sites do not differ from the rocky habitations their species select away from caves.

There are only two types of birds that are known to nest regularly in the dark zone of caves. The most numerous are the streamlined swiftlets which nest in caverns along coastlines from India eastward to the Philippines. These birds generally build their nests of seaweed, mosses, or feathers, which they bind together with their saliva. But one notable species, found mainly in Java, makes its nest entirely of saliva. Epicures pay high prices for the nests, which weigh less than a quarter of an ounce apiece: it takes three or four to make a bowl of bird's-nest soup, a delicacy that has been esteemed in China for thousands of years.

Huge populations of these tiny birds have established their homes in some caves. High on the cave wall or in ceiling crevices they deposit successive layers of a sticky, sweet-smelling substance that during the mating season flows from glands under their tongues. This substance hardens quickly when exposed to the air. When the nest is complete, the bird lays two tiny eggs inside.

Three or four times a year, nest-hunting natives invade these caves with blazing torches, ropes, and rattan ladders. They mount precariously to heights of a hundred feet and harvest the nests by tens of thousands. Birds whose nests have been taken will rarely build a second time. Natives returning again must be content with the feather-and-saliva nests of another species.

Alvin Novick, a Yale University biologist, has shown that swiftlets nesting in total darkness avoid obstructions and collisions by using a well-developed sonar system. In flight they make rapid clicks with their tongues and depend on the echoes to warn them of obstacles ahead.

Another bird that nests in caves is the oilbird *Steatornis caripensis* of Trinidad and South America. A relative of our whippoorwill, the "guacharo," as it is known locally, rears its young in the cave and flies out at night to feed on seeds and fruit.

Alexander von Humboldt, the famous German naturalist, was the first to study the oilbirds. When in 1799 he entered the Guacharo Caves near Caripe, Venezuela, he marveled at the dense growth of colorless tree sprouts on the cave floor. They had germinated from seeds dropped by the guacharos. The food stored in seeds permits them to grow

A phantom forest springs up from palm and laurel seeds dropped by oilbirds in the twilight zone of a South American cave. Because these two-foot seedlings receive a little light, they have a few pale green leaves and bend toward the light source at the cave entrance instead of standing straight up. Beyond the twilight zone such seedlings would be leafless and colorless and would eventually die from lack of light.

Oilbirds huddle on a cave ledge, their eyes fiery with light from the camera's flash unit.

Despite a three-foot wingspan, oilbirds maneuver without mishap through a cave's darkness. They avoid collisions by making sharp clicking sounds and listening for echoes. Once they leave the cave, however, these night fliers fall strangely silent and depend on keen eyesight to find their way through the darkened jungle.

for a time, and most of the seedlings were two feet tall before their food supply ran out. Von Humboldt was struck by the absolute whiteness of these doomed "forests" and by their rigid, undeviating verticality, but he knew that without sunlight they could not survive.

Von Humboldt was impressed by the din produced by the oilbirds. So was speleologist Donald R. Griffin and his party when they visited the Guacharo Caves in 1953 to see if the birds were actually flying in total darkness. Tests with fast panchromatic film proved the total absence of light. A cathode-ray oscilloscope showed that all the sounds were within the range of human ears. At dusk all sounds ceased, except for a steady stream of the sharpest imaginable clicks. Their frequency was between 6000 and 10,000 cycles per second, about four to five octaves above middle C. Like bats, the oilbirds guide themselves by the echoes produced by these clicks, but the clicks are not in the ultrasonic frequency range used by bats.

Solving the mysteries of bat flight

Bats possess the most highly developed sonar system. In absolute darkness they can fly without a blunder through twisting cave corridors or through the darkest forest. They can make quick U-turns when necessary and avoid obstacles, even wires, in their path.

No other night flier can locate flying insects with as much success. Swiftlets, nighthawks, and whippoorwills—all insect-eating birds—feed chiefly in the twilight, when they can see their prey. But most bats are nocturnal. They fly blind and rely on echoes, ranging in on each tiny mosquito or gnat with deadly accuracy.

The details of the bat's sonar system were discovered only recently, but as early as 1793 an Italian scientist, Lazzaro Spallanzani, suspected that bats guided themselves by hearing. He proved that bats which had been blinded could avoid obstacles and track down flying insects. But even those with sight were nearly helpless in the laboratory once their ears were plugged. They bumped into walls or flew head-on into wires stretched across the room.

In 1938 Donald R. Griffin, then a biology student at Harvard College, began a systematic study of the bat's means of navigation. Enlisting the aid of Harvard physicist G. W. Pierce, who provided a "sonic detector" that recorded very high frequency sound, he was able to show that in addition to certain audible squeaks, bats in flight gave out short bursts of ultrasonic sound.

Proof was still lacking that the echoes of their ultrasonic calls actually got back to the bats and that the bats could hear them. Scientists inserted recording electrodes in the inner ears of a number of etherized bats and hooked the electrodes to a complicated recording apparatus which included an oscilloscope. Recordings established beyond doubt that waves of high frequency do enter the animals' ears and set up impulses there. Later studies with active echolocating bats confirmed the findings and showed that the bats hear the echoes of their ultrasonic cries.

A perplexing question remained: How could the bat, while sending out a steady barrage of sound, hear the echoes well enough to guide itself by them? The oscilloscope pattern shows that each sound a bat makes is a separate, incredibly short note, lasting less than one-thousandth of a second. The bat hears and evaluates each returning echo in the brief interval between two outgoing notes. No wonder it can dodge a broom swung at it by a frightened householder.

A bat "sees" in the dark with its ears. The flying bat emits a steady stream of ultrasonic squeaks. When outgoing sound waves (*black*) hit an insect or other object, the echo (*brown*) bounces back. The bat pinpoints the exact location of an obstacle just as a sonar device does, by measuring the delay between the outgoing sound and its echo.

How good is bat echolocation?

Bats pay attention to their echolocation system only intermittently. When they are flying down familiar cave corridors, they have no need of fresh information, and so they evidently disregard the echoes. Explorers crawling through narrow cave passages are sometimes almost engulfed by bats. Sometimes, when an obstruction is suddenly placed along the route the bats customarily follow, a few in the front of the group collide with the unexpected object. But the signals they emit upon collision alert the bats behind them. Apparently they all tune in their sonar again and find a detour around the obstruction.

A bat's efficient sonar system makes it a difficult animal to

110

trap. Biologists who wish to study the migrations of bats often install *mist nets* in cave entrances in hopes of being able to catch and band them. Placed deep down a narrow, throatlike passage, these meshes of delicately fine hair succeed in tangling a few early comers. But their cries quickly alarm those behind, and they promptly whirl around and escape. If there is a hole in the net, the alerted bats find it immediately and dart through at full speed.

How bats catch insects

The efficiency with which bats detect and capture their insect prey was never suspected until recently. For years scientists had watched the characteristic diving, veering flight of bats, but it was not until 1953 that anything was known about how effective these maneuvers are.

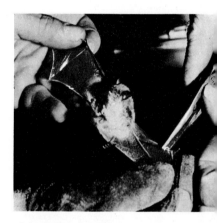

For three months Edwin Gould, now of Johns Hopkins University, methodically shot foraging bats at regular intervals each evening following their exit from their summertime roost in a church in Cape Cod, Massachusetts. He examined the stomach contents of hundreds of specimens. Nearly every bat he dissected had succeeded in filling its stomach within an hour or two. Gould discovered that it took the equivalent of 66 moths or 5000 gnats or midges to fill a bat's stomach. On the average, these bats pursued eight insects each minute and caught about six of them—an amazing batting average!

Scientists trace the movements of bats by clamping numbered aluminum bands to their wings. Because bats are very difficult to catch, most banding is done while they occupy hibernating sites in winter. Anyone finding a banded bat should report the band number, location, and date to the United States Fish and Wildlife Service, Washington, D.C.

During this same summer of 1953, Donald R. Griffin investigated another aspect of bats' extraordinary ability to capture prey while flying. He set up microphones and recording apparatus at a Cape God golf driving range where bats preyed upon insects attracted by the floodlights.

Griffin found that the bats had a general pattern of operating their sonar system. As a bat cruised about at random, seeking its prey, it emitted about 25 supersonic calls a second. When it detected an echo and swerved in pursuit, it stepped up the rate of its calls. In the last tenth of a second, when it closed in for the kill, it was sending out 200 to 300 notes per second. Griffin called the very fast final set of notes the "buzz."

The bat maneuvers so fast that observers have never actually observed the capture. But Griffin thought it highly probable that the buzz would not have ended so abruptly unless the bat had been successful.

Details of the capture were finally revealed in 1960, when Frederic Webster, of the Sensory Systems Laboratory at Cambridge, Massachusetts, perfected a camera and flash-gun combination that takes 1000 pictures per second. Webster taught bats of several species to catch meal worms shot up in front of the camera. The pictures reveal the bats' astonishing maneuverability. Once the bats were trained, hardly a meal worm got away. The biggest surprise was that a bat rarely caught an insect in its mouth. Generally

This unusual series of high-speed photographs, taken by Frederic Webster, catches a bat in the act of snatching a meal worm tossed up in front of the camera. The color patches on the key below indicate the relative positions of bat and insect during each exposure.

1 The bat locates the "flying" meal worm in the dark room.
2 The bat homes in on its target . . .
3 . . . and nets the insect in its wing membrane.
4 With a flip of its body, the bat scoops the worm into the outspread membrane between its hind legs.
5 Doubled over, the bat snatches the worm from the pocket formed by its tail membrane.
6 Seeming to tumble but still in full control, the bat flutters off with the meal worm clenched between its teeth.

the bat used its wing like a net to catch the meal worm and sweep it close enough to its mouth to grab it. Sometimes the bat plunged straight forward, scooping the worm up in its tail membrane and doubling its head downward to devour it.

How often the bat uses these maneuvers when it is hunting outside, nobody knows. But the fact that it can deploy tail and wings so swiftly and accurately accounts for its extraordinary success as a hunter.

Hiding by day in its snug nest, the pack rat slips from the cave at night to forage for berries, nuts, and other food. Since it may occupy the same nest year after year, its heap of trash may eventually grow three feet high. Besides food and material for its nest, the pack rat collects any shiny or colorful objects that it can find. A long-occupied nesting area may contain an impressive "museum" of pilfered objects.

Another cave visitor, the pack rat

Like most nocturnal animals, the pack rat *Neotoma* secludes itself in dark places during daylight and hunts for food at night. In woodlands it may choose a hollow tree for its nest site. In the desert, where it feeds on cactus plants, its nest of sticks is often covered or surrounded by needle-sharp cactus spines, which are a menace to marauders. But almost anywhere that the pack rat can find a cave, it lives underground.

Each night, so long as the weather permits, it leaves the cave to search for a supply of food, which it stores for later consumption. Possibly of even greater interest to it than food are the miscellaneous trinkets left in or near the cave by human visitors. Few objects that can be carried or dragged into its "museum" are overlooked.

During repeated visits to caves in central Pennsylvania author Charles Mohr became well acquainted with pack rats. As a rule he arrived about dusk and spent the night watching them. But he eventually decided that a full day spent in studying their comings and goings would be profitable. So he arrived about daybreak and set out an offering of dog food, sardines, crackers, and other food that he knew would be welcome. Then he trained his camera on the banquet spread for his guests and waited for them to appear.

But they failed to arrive. Hour followed hour, but the pack rats stayed out of sight. He could hear small tantalizing indications of their presence. From time to time they would shuffle sticks and stones. Anyone who was not aware of the pack rats' presence might have supposed ghosts were in the cave.

Dusk had fallen outside before the rodents made their first appearance. Then they came directly to the bait, quickly seized chunks of food in their forepaws, and darted back into the darkness with it. They paid no attention to the brilliant but brief electronic flash nor to the flame of the carbide lamp on Mohr's hat.

The pack rats' busy nighttime schedule took them out of Mohr's section of the cave for varying periods. To conserve his flashlight batteries, Mohr depended on a couple of big plumbers' candles and the pleasant light from the big carbide lamp on his helmet. But the extra weight of the lamp, with its four-inch reflector, makes such a hat a bit uncomfortable. So he put it down on the floor, with its lamp still burning.

Nursing babies, usually two or three to a litter, cling tightly to the mother pack rat. If she becomes alarmed and scurries away while they are feeding, the squirming young ones sometimes are dragged helplessly along.

A few minutes passed. All at once a pack rat appeared at the far side of the cavern, darted forward across the floor, and ran headlong into the flaming lamp.

The little animal must have singed its whiskers. It backed off in confusion, sniffed the air a couple of times, then made a wide detour around the helmet. Anyone who has followed a familiar route through a dark room at home will understand how startled the pack rat was because of an unexpected shift of furniture.

How does the pack rat navigate in darkness?

The pack rat, like most night animals, follows established routes through known territory. They are not learned in one easy lesson. Cautious, gradual exploration teaches the pack rat where there are dead leaves, a dangerous deep crevice, or obstacles such as fallen stalactites. These landmarks give it its bearings, and the repeated pattern of its muscular movements probably help in familiarizing it with the route.

Like the bats, when proceeding on a path it already knows, the pack rat disregards its detection systems, even its eyes. Sight at all times is a secondary sense. Moving in the cave's absolute darkness, or in the dim night outside, the pack rat finds its food by smell and touch. And unquestionably, it is guided by its own animal scent and the drops of urine left on its well-traveled paths.

Early observers of cave life decided that the pack rat was

blind, but there is no anatomical evidence to support this belief. The pack rat's eyes have the same structure as those of a deer mouse or flying squirrel. They are proportionally smaller, but perfectly functional. They have not degenerated because the pack rat continually moves in and out of the cave's absolute darkness.

What animals are potential cave dwellers?

If a landslide should seal the entrance, could the pack rat survive and reproduce as a full-time cave animal? It could, but only if food continually fell in or was washed in, if some of its spare nests could be dismantled and used in building new ones—and if it did not become bored to death with the same collection of trophies! But it is not likely that all these conditions would be met, and it is particularly unlikely that there would be enough food.

A potential cave dweller cannot count on a large food supply. That is why no birds or mammals spend their entire lives in caves; both groups must eat more or less continuously to maintain their high activity and body temperatures. But since reptiles do not have high body temperatures, they do not need so much food. You might think that they would make good candidates for life in caves.

Snakes might be able to survive in caves. They go a long time between meals, and some, the rattlesnakes and boas, have heat-sensing organs that enable them to catch mam-

This copperhead regularly sought shelter from the midday summer sun in a cave entrance. Rattlesnakes, too, have been found in cave entrances, but snakes are seldom seen in the cool cave interior where food is scarce.

Cave animals are classified according to their degree of adaptation to cave life. Troglobites live in caves and nowhere else. Troglophiles can complete their life cycles either in caves or in suitable habitats aboveground. Trogloxenes regularly enter caves, but they must return periodically to the surface.

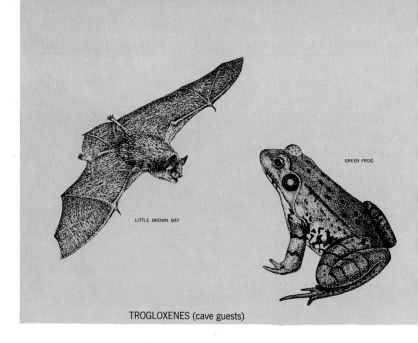

GREEN FROG

LITTLE BROWN BAT

TROGLOXENES (cave guests)

mals or birds in the dark. But even these snakes need a big meal occasionally and so could live only in a cave with a big bat colony. Frogs and salamanders are more likely to survive in caves because their *metabolic rates* (the rates at which they break down foods to produce energy and living tissues) are lower than those of any reptiles. Fish, some spiders and their relatives, crustaceans, and insects are good potential cave dwellers for the same reason.

Of the animals that have low enough metabolic rates, some are more likely to colonize caves than others. Frogs are poor bets because they cannot complete their life history in the cave. In White Oak Sink in Great Smoky Mountains National Park, one of the authors saw fat wood frogs that had apparently fallen unharmed into the sinkhole and were thriving on the insects they found in the cave. He found no tadpoles, however. Tadpoles eat green algae, and green plants do not grow in the darkness of caves.

Something more than a low metabolic rate is necessary for life in caves. Potential cave dwellers must also have the sense organs and behavior necessary to find food and to reproduce in caves; such animals are said to be *preadapted* to cave life. But not all of them are likely to colonize caves.

The most likely colonizers already live in cavelike environments. These are insects that live in leaf litter, crustaceans that are nocturnal and live under rocks, and fish that live in springs. They often wander in and out of nearby caves. Animals that are adapted to life in environments similar to caves and are also regular cave visitors are called *troglophiles*, "cave lovers." Full-time cave animals are called *troglobites*, "cave

118

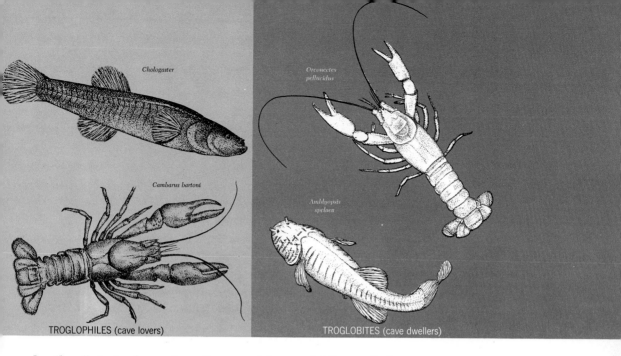

Cholaster

Cambarus bartoni

Orconectes pellucidus

Amblyopsis spelaea

TROGLOPHILES (cave lovers)   TROGLOBITES (cave dwellers)

dwellers." Animals, such as frogs and bats, which regularly visit caves but cannot complete their cycles underground are called *trogloxenes*, "cave guests."

A crayfish hunt in Bronson–Donaldson Cave

Crayfish are good candidates for cave existence. All of them could become troglophiles in the sense that they possess the organs and the low metabolic rates necessary for survival in caves, and, indeed, many have become troglobites. In some places both troglophilic and troglobitic crayfish may be found in the same stream. But can they coexist inside a cave? You can answer this question by visiting Spring Mill State Park in southern Indiana.

A map of Spring Mill State Park shows that Upper Twin Cave is the source of a stream that flows across an open area before entering Lower Twin Cave. After emerging again briefly, it passes underground through Bronson–Donaldson Cave. Then it reappears as a surface stream and flows through a deep valley and finally into a lake. If you walk along the lakeshore after dark, you should find crayfish quite easily. There! Your light reveals two bright spots, about half an inch apart! Keep the light on them. They are the eyes of a crayfish, shining, like a cat's eyes, by reflected light.

If you want to catch this crayfish, it is wise to prod it with a stick. This distracts it while you reach down and grab it just behind its big, powerful pinching claws. Its size is impressive; it has a thick body and is over five inches long.

There are surprisingly many crayfish. They seem to be everywhere. With the permission of the park naturalist, you catch, measure, and release twenty-five in half an hour.

In daylight you can walk up the stream valley from the lake to the large, nicely arched entrance of Bronson–Donaldson Cave. The entire valley was once a cave, but now the only signs of the former roof are big blocks of limestone covered with mosses and ferns.

From the wooden platform at the cave entrance you can hear the loud rumbling of the stream rushing out of the cave. The park naturalist explains that to tour the cave you must travel by boat. The special tour boat is almost as narrow as a canoe: a wider boat could not navigate the narrow gorge just inside the cave. The noise is louder as the park naturalist paddles the boat into a small room about sixty feet inside, but you discover that the thirty-foot waterfall you imagined is only a small, four-foot cascade. Here the naturalist inflates a rubber life raft which will carry you over pools of deep water as you travel upstream through the cave. While he is busy, you put on rubber hip boots and then look for crayfish. There's one now, climbing up the cascade. It has eyes; it is brown. Very likely all the eyed crayfish in the cave have followed this route. How many more will you find?

For 4200 feet a succession of sewerlike tubes, piles of debris, and huge boulder-littered rooms stretch before you. The cave passage follows a zigzag course. At the sharp bends are pools, deep and quiet, and between them is shallow swift-flowing water. Now and then you get out and wade beside the boat. Eventually you step into a pool too deep for your boots. The first shock of cold water takes your breath away, and you seem to be paralyzed from the waist down. But the water in your boots soon warms—from your own body heat.

Your discomfort is forgotten as you spot a second crayfish, this time a white blind troglobitic one. Then you see more

A rubber raft is indispensable for crossing the eight- to twelve-foot-deep pools in Bronson–Donaldson Cave. Indiana State Park Service officials permit only scientists engaged in research to explore beyond the tourist portion of the cave.

A pale shadow of its surface-dwelling relative, the blind and colorless cave crayfish (*top*) seems in every way more delicate than its larger surface relative (*bottom*). The cave species is slower-moving and able to fast between infrequent meals. It has a slimmer body, more slender legs, and longer antennae than the dark, robust surface dweller.

120

The large compound eye of a surface crayfish (*above*) glints with hundreds of minute lenses. The troglobitic crayfish (*opposite page*) has no eye, only a knoblike supporting stalk.

and more of them—dozens, scores. At the end of six exciting but exhausting hours you approach the upstream entrance and pause to figure out how many animals you have seen: 120 blind white troglobitic crayfish, but only 7 eyed brown troglophiles. Why so few? Apparently they cannot compete with the troglobitic forms where food supply is poor.

In the twilight zone of the upstream entrance to the cave, you suddenly realize that you haven't seen any white crayfish for a few minutes, but you've checked off two more brown ones. It appears that the troglophiles are outcompeting the troglobites in this area. A quick look just outside the cave confirms your suspicion: in a few minutes you check off fifteen more troglophiles. They seem to be more numerous than the troglobites in the cave.

You've come a long way from the lake below Bronson–Donaldson Cave. Maybe these aren't the same kind of crayfish you saw in the lake. They don't look different. Measure some and see.

Yes, they are slightly different. Their antennae are longer, and their legs and bodies are more slender; but they are nowhere near as slender as the white troglobitic crayfish— *Orconectes pellucidus* they are called—in the cave. However, the eyes of these *Cambarus bartoni*—the crayfish you have

122

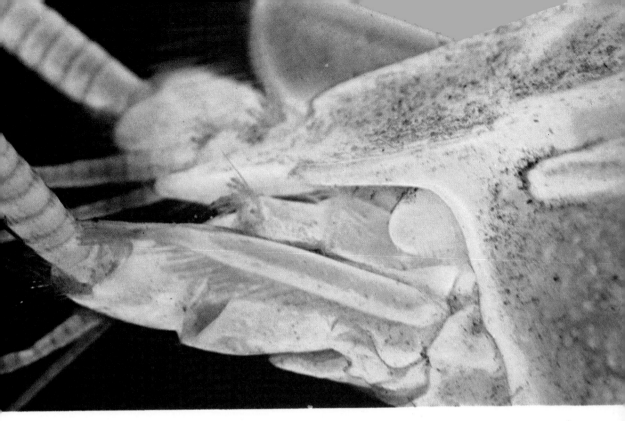

just found outside the cave—are no smaller than those of the lake dwellers, and these crayfish are just as dark brown. Why? Perhaps it is because most of the population lives outside of the cave, where their eyes are useful and where their dark color serves to camouflage them. Predators find it easier to catch crayfish that have smaller eyes or lighter pigment, and these crayfish do not survive. Evidently, though, there is no such disadvantage in having longer legs and longer antennae. Crayfish with these traits must have an advantage in the twilight zone and further back in the cave.

Although this cave crayfish has no eyes, it retains eyestalks since the structures contain several organs not related to vision.

If there were no troglobitic *Orconectes* in the cave already, these troglophiles would be more common there. But food is so scarce inside the cave that the larger troglophiles cannot compete with the smaller troglobites, which need less food and are more efficient at finding it. If *Cambarus* were trapped in a cave without *Orconectes*, it would slowly turn white and become blind over many thousands of generations. It would get slimmer, and its antennae would lengthen. But until *Cambarus* enters a virgin cave, it will not change from its present troglophilic state.

To see the next stages of adaptation to permanent life in caves, you must look for troglophiles living in caves without troglobitic competitors. The springfish is a good example.

123

The springfish

Not strictly a surface dweller yet not entirely adapted to cave life, the springfish, *Chologaster*, manages to exploit both worlds. It is common in certain surface springs, where it hides by day and hunts after dark; it also enters caves and occasionally even breeds in underground streams.

In southern Illinois, water wells up at the base of the limestone cliffs that rise above the flood plains of the Mississippi and Big Muddy Rivers. The spring water trickles over broken rocks and beds of water cress for a few hundred feet before entering the swamps formed by oxbows, portions of the old river channel now cut off from the main stream. In some of these springs, according to rumor, fish appear from underground. Guessing that they might be troglophiles, author Thomas Poulson decided to investigate.

Visiting the area by day, he found no fish, but there was a multitude of amphipods, as many as 500 per square foot of stream bottom. And flatworms were present in great numbers too. What a bonanza for hungry fish—but where were the fish?

On a hunch that they were nocturnal, Poulson returned after dark and started looking for them in the stream, with no luck. Then his probing flashlight spotted them in the spring source at the base of the cliff. They were at every level, from the surface to the bottom of the pool. There were about thirty fish, hanging in the water at all angles. Most of them, the adults, were about three inches long; others, only half an inch long, were the young of the year. Their small eyes and brown color were an important clue to their identity: *Chologaster agassizi*, the springfish. Though they can be seen aboveground throughout the warmer part of the year, they go underground in winter and breed there.

Can these small-eyed fish live permanently in total darkness? Evidently they can. At Jewel Cave, Tennessee, *Chologaster* lives well inside the cave, but it is also seen regularly in sunlit springs near the cave. *Chologaster* is a troglophile, able to come and go. Since it is nocturnal at all times, it can find food in the dark. And it has already overcome the mysterious limitations that prevent most animals from breeding successfully in caves.

Biologists are now studying these transitional troglophiles. By learning more and more about them, they hope to discover the steps by which surface dwellers evolved into the highly adapted troglobites able to live in caves and nowhere else.

Visitors to Mammoth Cave National Park can take a boat ride down Echo River, which lies 360 feet below the surface. During spring floods, the usually placid Echo River becomes a raging torrent bringing food and springfish into the cave from the Green River outside.

Two cave fish compared

If you descend to the River Styx in Mammoth Cave after spring floodwaters recede, you may find the troglophile *Chologaster* in the same pool with the troglobitic fish *Typhlichthys subterraneus*. During floods *Typhlichthys* is sometimes carried to the river from farther back in the cave.

Since the River Styx becomes a side channel of the surface Green River during floods and so has plentiful food, *Chologaster* can survive there and outcompete *Typhlichthys*. But life is precarious in the River Styx. In years when the Green River does not flood severely, the *Chologaster* in River Styx starve to death, and they will not appear there again until they are washed in from outside springs during some future flood.

When author Poulson and his companions went looking for fish in the River Styx, conditions were excellent. The spring floods had been the greatest in the history of Mammoth Cave—sixty feet above low-water level! The boardwalk was covered by inches of sand, silt, and fine organic debris deposited by the receding floodwaters. They spotted both a brown and a white fish in the first pool and climbed down to look more closely.

The differences between the troglobite and the troglophile

were obvious when they watched them swim side by side. With its almost colorless body tapering back from a large head, and with long, graceful fins, *Typhlichthys* can be told at a glance from the brown cigar-shaped *Chologaster*, with its short, stubby fins. Their swimming behavior is different too.

Chologaster swims for a bit, rests, and swims again. The fish heads in no particular direction and moves barely four or five body lengths per minute. Its fin strokes are quite short and ineffective: each one moves the fish only a third of its body length. *Typhlichthys* is clearly more efficient. With one stroke of its long, powerful fins, it glides nearly a full body length. It pauses for shorter periods than *Chologaster* and covers eight to ten body lengths per minute. Since it turns less often, it covers more territory with less effort.

When Poulson and a companion tried to catch one of each for laboratory study, they noted further differences. Before they got their net near *Chologaster*, it darted away wildly, without direction. It was easy to get close to *Typhlichthys*, but the fish was tantalizingly hard to catch. It easily kept a little way ahead of the net, diving straight for the bottom every time. In contrast to *Chologaster*, it seemed to know where the net was coming from.

You must have special permission from the National Park

Gliding slowly through the water, the blind cave fish *Typhlichthys* pauses occasionally to nose under rock ledges in search of isopods and other food. Vibration receptors on its head and sides detect movements in the water and guide it to its prey.

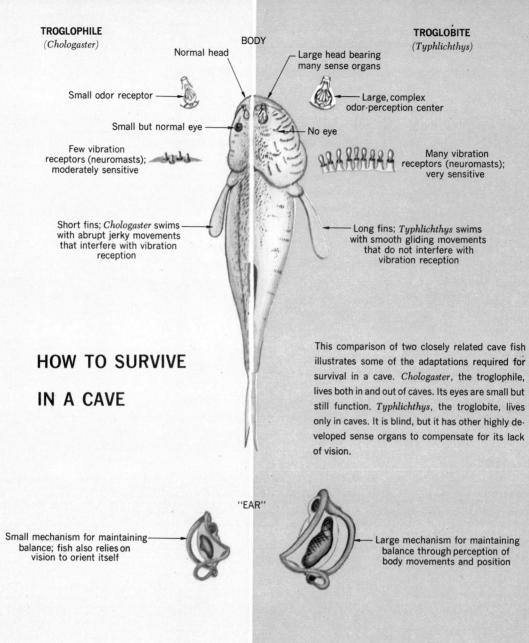

TROGLOPHILE
(*Chologaster*)

BODY

TROGLOBITE
(*Typhlichthys*)

Normal head

Large head bearing many sense organs

Small odor receptor

Large, complex odor-perception center

Small but normal eye

No eye

Few vibration receptors (neuromasts); moderately sensitive

Many vibration receptors (neuromasts); very sensitive

Short fins; *Chologaster* swims with abrupt jerky movements that interfere with vibration reception

Long fins; *Typhlichthys* swims with smooth gliding movements that do not interfere with vibration reception

HOW TO SURVIVE
IN A CAVE

This comparison of two closely related cave fish illustrates some of the adaptations required for survival in a cave. *Chologaster*, the troglophile, lives both in and out of caves. Its eyes are small but still function. *Typhlichthys*, the troglobite, lives only in caves. It is blind, but it has other highly developed sense organs to compensate for its lack of vision.

"EAR"

Small mechanism for maintaining balance; fish also relies on vision to orient itself

Large mechanism for maintaining balance through perception of body movements and position

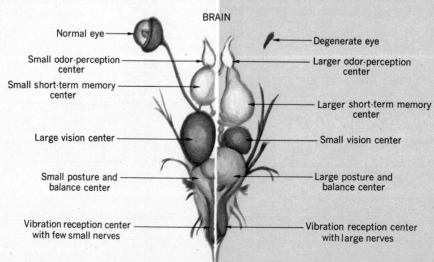

BRAIN

Normal eye

Degenerate eye

Small odor-perception center

Larger odor-perception center

Small short-term memory center

Larger short-term memory center

Large vision center

Small vision center

Small posture and balance center

Large posture and balance center

Vibration reception center with few small nerves

Vibration reception center with large nerves

Service to study fish in Mammoth Cave, but any visitor can easily observe the differences between the two fish by stopping at the display tanks along the tourist trail. *Chologaster* is brown and has dark lines on its body. Its eyes are small but normal in appearance. Where *Typhlichthys'* eyes should be, however, there are only bulges, and the fish has just the faintest speckling of pigment on its body. Its head is larger than *Chologaster's* and, like the sides of its body, is covered by prominent ridges, the *lateral line system*. All fish and amphibians have this system, but it is not often visible.

If you anesthetize the two fish in a laboratory and place them in a dish of water under a low-power microscope, you can observe that the ridges on *Typhlichthys'* head are actually rows of the lateral-line sense organs, the *neuromasts*. You can also see that *Chologaster* has similar rows of neuromasts. Each neuromast has a gelatinous rod—a *cupula*—that sticks out into the water.

You can watch this lateral line organ work if you attach a live water flea to a pin with a bit of petroleum jelly and move it toward one of *Typhlichthys'* neuromasts. Using the microscope, you will observe the water flea flailing its tiny antennae as it tries to swim, thereby creating minute water movements which move the cupula. *Chologaster's* cupulae respond only when the water flea is half an inch away, but those of *Typhlichthys* respond when it is as far as an inch and a half away. *Typhlichthys* has four times as many neuromasts as *Chologaster*, and they are more than twice as sensitive to changes in water movement. It is no wonder that *Typhlichthys* survives better than *Chologaster* where food is scarce. This sensitivity also explains why *Typhlichthys* escaped the author's net more efficiently.

Now that you know how the lateral line works, you may also see an advantage to *Typhlichthys'* way of swimming. Its smooth, gliding movement creates no agitation in the water that would reduce the effectiveness of its lateral line in detecting either prey or obstacles. The neuromasts are not "confused" by the fish's own movements.

Its almost effortless manner of swimming benefits *Typhlichthys* in another way: *Typhlichthys* can swim farther than *Chologaster* with lower fuel consumption. In effect, it gets more miles to the gallon. Yet *Typhlichthys* does not sacrifice speed for efficiency: it can swim just as fast as *Chologaster* can when the occasion demands.

Several highly adapted species of cave salamanders found in Texas probably looked somewhat like this common surface dweller before they entered caves. Like its cave-dwelling relatives, this salamander retains feathery gills and other larval characteristics even as an adult.

An unusual cave salamander

In central Texas, the salamander *Eurycea rathbuni* had been living in deep underground waters for uncounted millennia. It had never been seen by human beings until 1895, when some artesian-well drillers at a fish hatchery near San Marcos sank a shaft 188 feet to tap waters in Purgatory Creek, an underground stream. Grotesquely slim white salamanders rose to the surface. They had no discernible eyes, and their tapering heads culminated in blunt, spoonlike snouts.

This salamander is of great interest to biologists. Its strange appearance is the result of a long evolutionary history. Remarkably, it is still possible to see how it became increasingly adapted to the cave environment. There are several other related species of the salamander *Eurycea* living in the surface waters and cave waters of central Texas, and they no doubt represent stages similar to those through which *Eurycea rathbuni* passed over tens of thousands of generations.

Isolated for so long in the food-poor waters of Purgatory Creek, this troglobitic salamander has accumulated some unusual traits. Like the cave fish, *Eurycea rathbuni* has a well-developed lateral line. You have seen how this system detects minute agitations in the water, thereby informing the hungry salamander whenever something, which is prob-

130

ably alive, is moving nearby. But since food is scarce, *Eurycea* must have energy-conserving adaptations, and it does. Long legs enable it to search wide areas with little effort—low metabolic cost—and help to keep its head high so that it can sense disturbances in the water over a wider area. Its slow, stealthy stalking motions ensure better hunting, and disturbance of the water is so slight that there is little interference with the function of its lateral line.

Food is so scarce in Purgatory Creek that *Eurycea rathbuni* is the only large predatory troglobite that can exist. Even nonpredatory troglobites cannot coexist with relatives unless the cave has abundant food. In a few cave areas, especially in the southern United States, there is apparently enough food, since several crayfish species have become isolated in the same cave. The results are most intriguing.

Two Texas salamanders, each inhabiting separate cave systems, probably resemble past stages in the evolution of the highly adapted *Eurycea rathbuni*. Compared with their surface relative *(opposite page)*, they show progressively greater loss of pigment, degeneration of eyes, elongation of legs, slimming of the body, and flattening of the snout. Possibly at some time in the future both species also will develop the grotesque modifications of *Eurycea rathbuni*. For a look at that rare and remarkable product of evolution, see the next two pages.

The discovery of a rare crayfish

Potential business rivals often divide up a market instead of struggling to monopolize it. Rather than waste energy in competition, each specializes in one area of the market and develops more efficient ways of exploiting it. In a similar way, when there is enough food in a cave, two closely related species can coexist there if they either exploit different sections of the habitat or specialize in eating different sorts of food. In 1940 Horton Hobbs discovered a situation in which both of these processes must have occurred.

Hobbs spent years observing and collecting cave crayfish in northern Florida. He found greater numbers and variety than were known in all the rest of North America. Collecting conditions were ideal. The continuous northern belt of limestone is exceptionally porous and soluble. Because of Florida's heavy rainfall, the limestone remains saturated, and the countless miles of uncharted caverns are filled with water

Entranceway to a unique laboratory of evolution, vine-draped Squirrel Chimney in northern Florida is a vertical shaft that plunges straight down to the water level sixty feet below.

Using safety ropes and wire ladders, scientists descend into the underworld at Squirrel Chimney. Alongside the deep pool at the bottom, an opening leads to water-filled passageways populated by two remarkable species of blind crayfish.

and with more food than exists in most caves. In many places the land has collapsed, forming hundreds of lakes and ponds; doubtless, many of them are interconnected. Sinkholes are unusually abundant, and many have narrow shafts that penetrate into water-filled caves thirty to a hundred feet below.

Hobbs visited some caves repeatedly because their fluctuating water levels often altered conditions, making it easy at one time, impossible at another, to reach the best collecting sites.

Squirrel Chimney, Hobbs recalled, was a cave that had shown promise of being a good collecting site when he visited it before. The sixty-foot smooth-sided vertical shaft dropped directly to an underground pool, where he had found fair numbers of a big white blind crayfish, *Procambarus pallidus*. The water surface covered only the space of a dinner table, but its unplumbed depth was obviously great.

135

A scuba diver on a crayfish hunt sets out to explore the flooded cave system at the bottom of Squirrel Chimney. The compass on his wrist and the reel of twine held by his companion will later guide him back to the entrance.

Only a diver could tell how extensive the underground lake might be.

Hobbs enlisted the help of student William MacLane, an expert swimmer, and headed for Squirrel Chimney. Descending their cable ladder, the explorers discovered the water level to be at a new low. Shoulder-high, as they stood on a narrow ledge at the bottom, was a windowlike opening. They squeezed through and found themselves standing on the brink of a deep, water-filled canyon. MacLane lowered himself into the water and skillfully netted half a dozen specimens of *Procambarus*. The smallest were over three inches long; many were nearly four inches. From time to time, he brought his trophies to Hobbs, who stored them in a water-filled jar.

Just as they were about to leave, MacLane caught a glimpse of something white floating toward the bottom from under a ledge. He scooped it up and, returning to Hobbs, handed it to him. Hobbs took one look and became very excited. The small white crayfish he was looking at was a type utterly unknown to science. It was barely an inch in length, but its remarkable slimness and unusually long legs and antennae immediately set it apart from any previously known species.

The younger man could hardly be persuaded to leave the cave. He spent several hours looking for more specimens, but not a single one could he find. Still it was a memorable day for both explorers. Hobbs named the new species *Troglocambarus maclanei* in recognition of MacLane's help.

Suspecting that the little crayfish might avoid the open stretches of water, Hobbs engaged a professional diver with a face mask and an underwater light to search channels where the water rose to the ceiling. The diver made a methodical survey of the underwater passages for an hour before he hit pay dirt. He rolled over on his back. Directly overhead were two of the dwarf crayfish hanging by their long, spidery legs from the ceiling. In the next forty-five minutes he found five additional *Troglocambarus* specimens. All were hanging upside down from the ceiling. They were far scarcer than the three- to four-inch-long *Procambarus*, but much easier to catch. The diver picked them from the ceiling. He told Hobbs it was as easy as picking grapes.

Apparently MacLane had discovered *Troglocambarus* by a lucky accident. The little crayfish probably was dislodged from the ceiling by ripples created by the swimmer.

A *history of competition*

The largest *Troglocambarus* known is just an inch and a half in length. These creatures are such lightweights that they can even hang from the water-surface film—quite a contrast to the three- to four-inch *Procambarus*, which do well to hold on to a rough sloping wall.

When Hobbs made detailed anatomical studies of both species, he found evidence that they are closely related. The differences between the two had grown steadily during their long coexistence underground. Their history, as author Thomas Poulson reconstructs it, is typical of cases where two similar species confined to the same habitat depend on a common food source.

If this theory is correct, the ancestors of *Troglocambarus* were in the cave first. A typical individual was probably over two inches long. Adults foraged mostly on the bottom and rocky sides of the cave, but the immature crayfish spent some of their time on ceilings and floating debris. The species was already beginning to develop metabolic economies.

Later the cave was invaded by ancestors of *Procambarus*. They were bigger, possibly almost three inches long. Their slight advantage in size and their greater vigor made them more effective feeders at the bottom of the pool, where food was more abundant. Any *Troglocambarus* that competed for food in this area probably fared badly and died.

Conditions on the bottom favored the largest *Procambarus* individuals. They got more food, lived longer, and produced more offspring than smaller individuals. This process of natural selection weeded out the smallest, weakest individuals in each successive generation until the species reached its present size of three to four inches.

On the other hand, the smallest *Troglocambarus* with slower rates of growth were especially favored. These lighter-weight individuals probably spent more time feeding on the ceiling and under ledges. They escaped the ruinous competition on the bottom of the pools that was eliminating the larger and hungrier members of their species. Eventually the whole population consisted of midgets.

Along with the changes in size of these two crayfish there had to be changes in feeding. *Procambarus* did not change habitat, so it probably just became a more efficient feeder. But *Troglocambarus* became much smaller and shifted habitat drastically. What happened to its feeding parts?

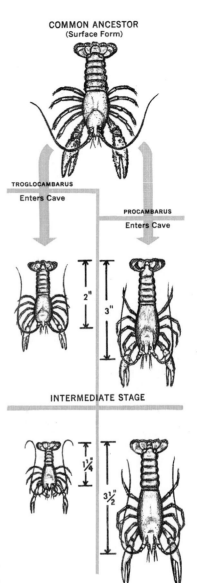

COMMON ANCESTOR
(Surface Form)

TROGLOCAMBARUS
Enters Cave

PROCAMBARUS
Enters Cave

INTERMEDIATE STAGE

MODERN FORM

Despite obvious differences, the two species of crayfish found in Squirrel Chimney probably evolved from a common ancestor. Scientists speculate that when larger *Procambarus* invaded caves already populated by *Troglocambarus*, the smaller crayfish evolved into a midget form that avoided competition with its larger relative by occupying a different part of the water-filled cave.

137

Antennae

Third maxillipeds

TROGLOCAMBARUS

How does Troglocambarus feed?

Even a fully grown *Troglocambarus* retains certain features usually found in young crayfish: relatively long legs and antennae and oversized mouth parts. The feeding apparatus, one of the most distinctive features, is especially interesting to biologists. The outermost appendages, the third maxillipeds, are equipped with closely interlocking hairs and are particularly large. What can their special use be?

At first Poulson conjectured that the crayfish, hanging upside down, might fan the pool with its tail and thus drive water forward through the third maxillipeds. The hairs would filter out any plankton in the water. Hours of observation by Martha Cooper, a graduate student at Yale University, have discredited this idea. *Troglocambarus* never fans its tail, except on those very rare occasions when it is obliged to swim.

On two separate occasions *Troglocambarus* has been seen striking out, with its third maxillipeds, at a fruit fly floating on the water. The movement resembled that of a praying mantis, which uses its spiny front legs to capture prey. But the crayfish did not catch the fly, and there is no certainty that it often feeds in this way.

Most of the time this crayfish hangs nearly motionless from a floating stick or the underside of a rock ledge, and if it slips off its perch, it slowly climbs or swims back. While hanging upside down, these little crayfish are perpetually "preening." They draw their legs and antennae repeatedly through the hairs on their large maxillipeds, thus clearing off whatever debris has settled on these sensitive appendages. It may well be that this is a way of feeding: their legs and antennae are covered with tiny hairs which continually accumulate debris, and most debris carries numerous bacteria and protozoans. Further observations should help to explain

Spidery *Troglocambarus (top)* habitually clings to the undersides of ledges and floating sticks. Never more than an inch and a half long, this translucent featherweight can even hang from the water-surface film. The hairy "legs" extended beneath its head are actually oversized feeding appendages. *Troglocambarus'* waxy three- to four-inch-long relative *Procambarus (bottom)*, like most crayfish, spends its life creeping slowly across the bottoms of pools. Notice its relatively small feeding appendages.

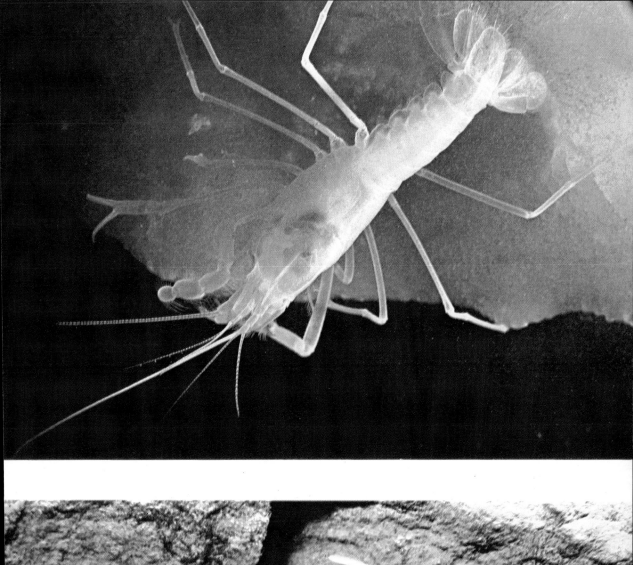

Continually cleaning its legs and antennae by drawing
them through the hairs on its feeding appendages,
Troglocambarus probably collects and eats microscopic
bacteria and protozoans that settle on its body.

the meaning of these actions. They are, in any event, amus-
ing to watch. The tiny crayfish has to use the claw on one of
its front legs to reach its antennae. It then takes five or six
determined tugs to draw such a long appendage all the way
through its maxillipeds.

Biologists spend many hours observing such details of
animal behavior. Only gradually can they form a clear picture of the animal's way of life and the work performed by
its organs and appendages. If the animals are small and slow-
moving, infinite patience and keen eyesight are needed.
Otherwise, even trained scientists can overlook important
evidence or jump to false conclusions.

While examining the crayfish collection in the United
States National Museum in Washington, Martha Cooper
opened a jar of specimens of *Procambarus* collected in 1890
in Sweet Gum Cave, Florida. To her astonishment a single
perfectly preserved little *Troglocambarus* was included. To
her, its tiny size, abnormally long legs and antennae, and
oversized maxillipeds shouted *Troglocambarus*, but for the
previous seventy-five years this unique product of evolution had been passed over as an immature specimen of the
larger species. Apparently, MacLane and Hobbs had not
been the first to find a *Troglocambarus*, but they were the
first to appreciate what they had found.

The fact that these two crayfish are often found together
adds further support to the hypothesis that *Troglocambarus*
evolved its way as a result of competition with *Procambarus*.
Did *Procambarus* change its feeding habits as a result of
competition?

Feeding habits of Procambarus

Procambarus did not change its feeding habits; it only became more efficient. If it had not done so, it would not have
survived in caves.

The feeding efficiency of an ordinary cave crayfish such as
Procambarus is evident if you compare it with an eyed sur-

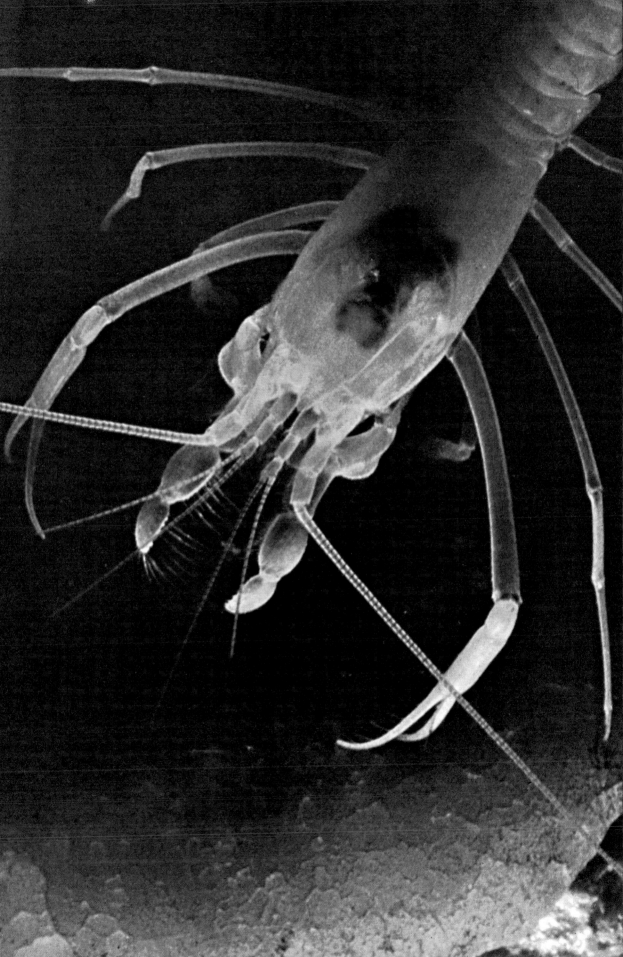

face relative. To make the comparison fair, you must "blind-fold" the surface crayfish and then put each crayfish in its own bowl together with a tiny injured worm.

First the cave species begins flicking its short, chemo-sensitive antennae, and then the ordinary crayfish begins also. They are both excited, turning this way and that while wildly moving their mouth parts. But the cave species moves more slowly and stops often. Once the crayfish get close to the worm, their behavior is even more different. The surface species walks back and forth, finally chances to walk over the worm, backs up, and grabs it. The cave species captures its worm much more neatly. It stays in one place, slowly moving its long touch-sensitive antennae from side to side. Suddenly the antennae stop; one is right above the struggling worm, and then the other swings around over the worm. After a while the crayfish walks directly to the worm and starts to eat it. So the cave species takes much longer to capture its prey, but it uses less energy. It senses its prey more quickly and locates it more precisely. This efficiency is a necessity in caves where food is scarce.

Beetles are found nearly everywhere

Of the million species of insects known to present-day biologists, about a third are beetles: Coleoptera. There are more species of beetles than of any other animal group. They include nearly all types of feeders: leaf eaters, sap suckers, grain chewers, wood borers, predators, scavengers, and parasites. These heavily armored insects have adjusted themselves successfully to all surface environments. There are beetles in the polar regions and the equatorial jungles, in deserts and rain forests, and even in metropolitan cities. Certain types of beetles are able to spend their whole lives in caves.

Beetles, like all the more highly developed insects, have a diversified life history. They regularly pass through four distinct stages. First eggs are deposited on a food-rich surface. Then *larvae*, or grubs, hatch from these eggs and feed for several months on the food at hand, periodically discarding the protective shells—*exoskeletons*—they have outgrown. In their third stage, as *pupae*, they do not feed. But inside the pupal covering, their immature tissues are broken down and reorganized. After this *metamorphosis*, they emerge from the

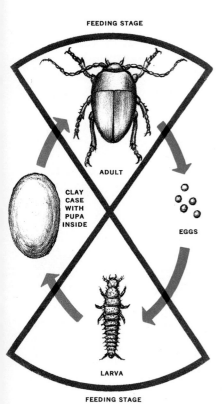

BATHYSCIOLA

FEEDING STAGE

ADULT

CLAY CASE WITH PUPA INSIDE

EGGS

LARVA

FEEDING STAGE

Two related beetles found in European caves show remarkable differences in their adaptations to food scarcity. *Bathysciola* inhabits caves where food is fairly plentiful. Its life cycle (*above*) resembles that of most beetles, with both adults and larvae actively foraging for food.

pupa stage as adult beetles, the fourth stage.

Such a diversified life history helps to explain the "biological success" of beetles. In a habitat where food is abundant, eggs can be scattered in widely separated places. The emerging larvae feed themselves and are able to develop into mature animals without help from their parents. The separate larval stage ensures full use of the food supply and eliminates competition between young and adults of the species.

But a food-rich area is never safe. Carnivores are on the lookout for eggs and larvae. At the critical moments when it is molting its exoskeleton, an insect is especially vulnerable. Many species of beetles lay their eggs in underground burrows which they stock with food supplies sufficient to feed the larvae during this immature stage. Others deposit their eggs in logs, in the forest litter, or under the bark of trees. These secretive habits increase the chance for survival.

Larvae of specialized cave beetles never feed

Beetle populations have been found in caves all over the world. Generally they inhabit the entrance or the twilight zone, where food is most concentrated. Caves that have a great deal of debris or have been heavily colonized by bats or crickets often support sizable populations of scavenger beetles, Catopidae and Dermestidae.

The females of these insects lay their eggs on fresh guano or vegetable and animal debris. There the larvae find enough food to provide the energy needed for growing, molting, and metamorphosis.

Several distinctive, highly specialized species of Catopid beetles have been found in the dark zone of food-poor caves. They are small, slim, pale, and totally blind. Their larvae go through a much simplified course of development in order to reach adulthood. It is this drastic simplification of their life history that enables these beetles to survive in the most impoverished habitats.

A Frenchwoman, biologist Sylvie Deleurance, studied two contrasting types of Catopids. Her account of the adaptations in life history that adjust each species to its habitat is a classic of cave literature.

Bathysciola schiodtei, the less adapted beetle, spends its life deep in bat caves or in debris-filled caves, where food is relatively plentiful. The female lays about twenty small yolk-

SPEONOMUS

FEEDING STAGE

ADULT

CLAY CASE WITH PUPA INSIDE

EGGS

LARVA

A compressed life cycle enables the beetle *Speonomus* to survive in areas where *Bathysciola* would starve. Only the adult searches for food. By laying fewer and larger eggs, *Speonomus* has practically eliminated the larval stage from its life history. Immediately after hatching, the fully developed larva encloses itself in a clay case, pupates, and transforms into an adult.

143

poor eggs each month, depositing them in piles of guano or litter. The larva hatches in about a week, then feeds for a period of seven to fifteen days. The next step is a critical one. To allow for growth, the insect must shed its tough, armorlike exoskeleton. But exposure of the larva's delicate skin to the air could kill it by drying. At this point the insect surrounds itself with a protective coat of clay, within which it molts. After its skin has hardened, the larva emerges from the case and feeds for another ten days. This feeding and molting sequence is carried on three times. After the fourth feeding period, the larva passes into the pupa stage, lasting two to three months. At the end of this fasting period, its metamorphosis complete, it emerges as a mature adult. During this cycle of from four to five months, the *Bathysciola* larva feeds or is otherwise active for forty days.

Speonomus longicornis, which lives in the darkness of food-scarce caves, could not survive such a life. Its larvae would have to travel long distances searching for the small, widely dispersed particles of food. Beetle larvae do not have very good legs for crawling; they would exhaust themselves and starve before they could locate and eat enough food to support their growth, successive moltings, and subsequent metamorphosis.

Two fresh-water shrimp illustrate a common adaptation of cave animals. The surface species *(below)* lays many small eggs. Once deposited, the incubating eggs are attached beneath the mother's body. The cave species *(opposite page)* lays only a few large-yolked eggs at a time. Thus the young cave shrimp will be relatively well developed when they have to find food for themselves.

Evolutionary change—which we shall consider in the next chapter—has adjusted the life processes of this species to its meager environment. The larvae of *Speonomus* do not feed. As soon as they hatch from the eggs, they encase themselves in clay and soon molt to the pupa stage. Only as adults are they required to forage in the cave. With their superior mobility and sense organs, adults can succeed where larvae inevitably would fail.

By foraging over wide areas, a female *Speonomus* accumulates a store of food well in excess of her needs. She lays very few eggs, two or three a month. But unlike the eggs of *Bathysciola*, they are tremendous, nearly a fifth of her own weight. Each egg has a huge food-rich yolk. The embryo remains within the egg, feeding and growing, for at least a month. When it emerges as a fat, fully developed larva, it is ready to molt instantly into the pupa stage. It has incorporated in its body enough material to carry on the tissue decomposition and reorganization that, in a period of seven to nine months, will transform it into a mature beetle.

Although *Speonomus* lays barely one-seventh as many eggs as *Bathysciola*, its larvae's chances of surviving are infinitely better. These eggs are rich in food. The larvae they contain cannot starve to death. They are able to develop directly into

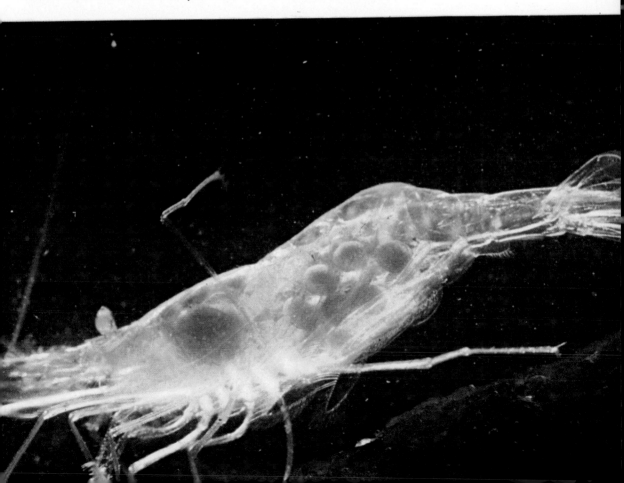

adults before they need to find food for themselves. In general, all the very specialized cave animals lay larger and fewer eggs than their relatives on the surface. This is one of the most universal responses to the perils of cave existence.

Prisoners in the world of darkness

You have seen that many animals make the hours of night their "day." A few can even navigate and feed in absolute darkness: owls hear their prey; bats and some birds home in on their prey by using echolocation; and pack rats feel and smell their way along. But even these animals cannot live underground permanently. They are all warm-blooded and therefore need more food than is ever found in caves. They have some of the sense organs necessary for life in caves, but not the essential low metabolic rates.

Many fish, amphibians, spiders, spider relatives, millipedes, crustaceans, and insects, however, are preadapted to cave life because they live in environments similar to caves. If they are often found in caves, we call them troglophiles; they have both the sense organs and the low metabolic rates necessary for spending their entire lives in caves. But they become permanent cave dwellers, or troglobites, only after they are isolated underground for millennia. During the course of their subterranean evolution they become better and better adapted to their dark, stable, food-deficient environment and less and less capable of survival in the food-rich, but extremely unstable and competitive, environment of the surface. The price paid by these specialized troglobites is high: they are prisoners within their own strange world and are forever prevented from returning to the world of light.

Blind but mysteriously sensitive to light, a cave-dwelling millipede curls helplessly in the glare of a flashlight beam. As it becomes more fully adapted to life in the cave over thousands of years, it may lose even this feeble response to the intrusion of light into its nighttime world.

Time and

Change

A remarkable salamander is found in caves in the Ozark Mountains of Missouri, Oklahoma, and Arkansas. This celebrated creature, the Ozark blind salamander (*Typhlotriton spelaeus*), was the first cave-dwelling amphibian found in America and attracted world-wide attention among scientists because the adults are blind and almost white. The immature salamanders, however, are so different from the adults that they might easily be mistaken for another species: they are neither blind nor white.

This Ozark salamander is born in the water and lives there until it transforms into a land creature—a normal procedure for most aquatic salamanders. But this procedure is not normal for troglobites. It is the changeover, the metamorphosis, that makes this salamander unique among cave salamanders. Fully cave-adapted salamanders do not transform: all retain their larval form throughout life, a condition called *neoteny*.

In its larval, aquatic stage, the Ozark blind salamander is chubby, its skin is pigmented, it carries on respiration through feathery red gills, and it has small black eyes. But as this salamander matures, it loses most of its pigment. Its gills disappear. Tiny blood vessels (capillaries) close to the surface

149

take over most of the respiratory function, and the red blood in them gives the skin a pink color. Eyelids develop, but they soon grow together. Beneath them, the eyes degenerate; they become reduced in size and have no further function. The head becomes quite angular, and the body and legs grow long and thin. The dark, chubby, eyed larva becomes a pale, thin, blind adult—capable of life out of the water.

Specialized, highly efficient sense organs enable this salamander to live in total darkness. In certain caves it is a permanent resident of the dark interior, producing young that generation after generation show cave-adapted characteristics.

In other apparently similar caves, young are never found. Why? Apparently the Ozark blind salamander lacks one decisive troglobitic feature: the capacity to survive on little food. The life it leads—its high metabolic rate, its active way of moving around, and its metamorphosis—demands a food supply greater than most caves provide. Metamorphosis is a particularly costly process requiring large amounts of food, and other cave-inhabiting salamanders have abandoned it. They simply live in the condition of neoteny, as "permanent larvae."

The Ozark blind salamander represents an important stage in the evolutionary chain that produced cave-restricted creatures. It cannot be called a troglophile: it does not breed outside. But it isn't a troglobite either. It is found only in caves where the food supply is greater than normal.

This salamander fares particularly well in bat caves, where it can feed on the isopods, amphipods, flatworms, spiders, and other small creatures that get their nourishment from the guano. In such a food-rich habitat, its whole life cycle is completed underground. But in caves where food is scarce, the larval salamanders must travel to the twilight zone or outside to find sufficient food. Often the adults, too, are found outside, returning to the cave only to breed. As you might expect, those individuals that live outside the cave still avoid light. They pass the daylight hours under rocks and emerge only after dark.

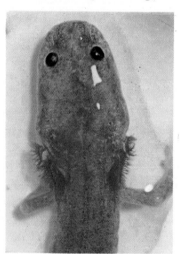

The two- to three-inch-long larva of the Ozark blind salamander lives in cave streams and pools but sometimes ventures aboveground. Unlike the adult, it has eyes, conspicuous coloration, and plumelike gills.

As the larval Ozark blind salamander approaches metamorphosis, it gradually loses pigmentation, its gills begin to regress, and its eyes become smaller in proportion to its head. Already losing the use of its eyes, the maturing larva begins to depend more on vibrations sensed through its lateral line system in order to locate isopods, flatworms, and other small prey.

150

Slender and ghostly pale, the fully transformed adult Ozark blind salamander is three to four inches long. Its gills have disappeared, and its eyelids have grown together over small, nonfunctional eyes. The adult salamander moves freely in and out of water and usually lives deeper in caves than the larval form.

Doomed to blindness

The discovery that the Ozark blind salamander is capable of a kind of Dr. Jekyll–Mr. Hyde existence, depending on the richness of its food supply, has helped to clear up an old misunderstanding. From reports that it might live outside of caves, G. Kingsley Noble in the late 1920s concluded that the deterioration of its eyes is caused by the lack of light in the cave environment and, therefore, is not hereditary. He thought that the salamander's eyelids would remain open and that its eyes would be unaffected if its metamorphosis occurred where light was present. At the time of his death in 1929, Noble was conducting experiments to see if an individual kept in the light retains vision in its adult stage.

Leon Stone of the Yale Medical School recently settled this question. As Noble had done, he watched the salamander transform and was able to follow the deterioration of its eyes. Stone found that during metamorphosis its eyelids grow together whether it lives in daylight or in total darkness. But eye degeneration starts even before metamorphosis. Indeed, the larval salamander begins to neglect visual clues. The right and left eyes develop at different rates. Stone recognized that the closing of the eyelids during metamorphosis might speed up the degeneration of the eyes, but he showed conclusively that it is not the cause of the degeneration. The loss of sight in this salamander is an inherited characteristic.

Adult Ozark blind salamanders can be said to be "doomed" to blindness. Scientists believe that it will be only a matter of time before the species acquires the additional adaptations possessed by more specialized troglobites, adaptations that add up to the ability to function on little food: metabolic economy we have called it.

Which animals have been in caves the longest?

The degree of eye and pigment degeneration in related cave species probably offers the best clue to the relative length of time they have been isolated underground. Those forms whose eyes have deteriorated farthest and whose pigmentation is most deficient are thought to have spent the longest time in caves. The argument is an important one.

Even nocturnal surface animals generally retain pigmentation and sight. During the day they hide; after dark they rely on senses other than sight. But in emergencies, it is helpful to see. Their coloration likewise is advantageous, making them less conspicuous to the predators that hunt them. Thus troglophiles have retained pigment and eyes. From time to time individuals are born with less pigment or with optical defects, but they are consequently more vulnerable when they venture outside the cave, and they thus leave fewer offspring.

Cave-dwelling animals, however, are not handicapped by their defective eyes. In the darkness, even those animals with good eyes cannot use them. A blind individual and one able to see are on the same footing. Provided their other senses are equally acute, each will fare well and produce offspring. By the same token, a pale individual will suffer no disadvantage. In the lightless cave, all animals are invisible.

Aboveground, such defects would be fatal for the indi-

Adult

Larva

OZARK BLIND SALAMANDERS

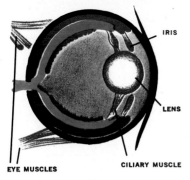

1 SURFACE-DWELLING FISH

IRIS

LENS

EYE MUSCLES

CILIARY MUSCLE

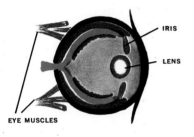

2 TROGLOPHILE *(Chologaster)*

IRIS

LENS

EYE MUSCLES

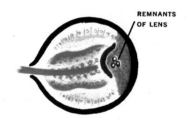

3 TROGLOBITE *(Typhlichthys)*

REMNANTS OF LENS

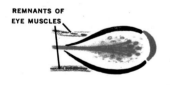

4 TROGLOBITE *(Amblyopsis)*

REMNANTS OF EYE MUSCLES

Focusing structures

Sensory structures (retina)

Supporting and protecting structures

Nutritive structures

ACTUAL SIZE OF EYES

1   2   3   4

viduals that possess them; none would long survive. But in a cave, the same defect will be preserved and transmitted through thousands of generations, spreading gradually through the whole population.

From time to time, further eye and pigment defects develop in certain individuals. So long as these new deviations from the norm do not reduce the animal's ability to survive, they will be passed on and in time become part of the inheritance of the entire population. The number of such hereditary changes, or *mutations*, that a species has accumulated gives scientists some idea of how long it has been in caves. For example, the two troglobitic fish species *Typhlichthys* and *Amblyopsis rosae* have eyes so reduced in size that they are visible only under a microscope. *Amblyopsis*, it seems, has been isolated longer. Its eyes are smaller, and more parts of the focusing apparatus, retina, and supporting structures have disappeared or are abnormal. *Typhlichthys* has lost certain other structures, such as eye muscles and some retinal pigment; but since it has not accumulated as many defects, we assume that it has not been a cave dweller as long.

Fossils in caves

Since we know very little about the rate at which living things are presently evolving, we can hardly predict with any certainty how long it may take for a given troglophile to become a troglobite. However, we can get some idea of the minimum or maximum time that it may have taken in the past for cave animals to evolve. We know, for instance, that

To see how the eyes of cave animals degenerate, compare the eye of a surface-dwelling fish (1) with the eyes of three cave fish (2,3,4). The cave fishes' eyes are smaller, and parts of them have degenerated or have disappeared altogether. The eye of the troglophile *Chologaster* (2), for example, has lost all traces of ciliary muscles. Otherwise *Chologaster*'s focusing mechanism is normal, and it still can see. The eyes of the two troglobites, on the other hand, are mere remnants embedded in tissue near the brain. *Typhlichthys*' eye (3) still has parts of a lens, but no eye muscles. *Amblyopsis*' eye (4) has lost the lens but retains traces of eye muscles. Although the eyes of the two troglobites have not degenerated in quite the same way, the result of their random and gradual accumulation of inherited defects is the same: both fish are completely blind.

ered so many well-preserved tusks, teeth, and bones that the American Museum of Natural History's noted paleontologist George Gaylord Simpson hurried from New York to direct the digging. The bones were those of peccaries, piglike animals, whose relatives still live today in the southwestern United States.

In all, more than a thousand skeletons were found, but none of them was complete. Bones and skulls were so thoroughly mixed that Simpson concluded that a whole herd must have drowned in some ancient flash flood and that their bodies then were washed into the cave.

When this happened is anybody's guess. There was no way to reconstruct the scene there in the Mississippi Valley when these peccaries were so astonishingly numerous. The handful of bones of a few other species provided no clues. They, too, were scattered at random, not preserved one on top of another in any sequence.

This peccary skull was among the thousands of bones found at Cherokee Cave.

Scraping carefully with a small blade in order to avoid damaging his find, a scientist cautiously removes the jawbone of a peccary from a layer of mud in Cherokee Cave, beneath St. Louis, Missouri.

Going up! A frogman displays his trophy, the fossil molar or grinding tooth of an extinct mastodon. Most of the remains recovered from Wakulla Cave are those of mastodons, but giant sloths and several other animals also died here thousands of years ago.

A scuba diver prepares to plunge into the depths of Florida's Wakulla Springs. Far beneath the surface is the hidden entrance to Wakulla Cave.

WATERY GRAVE AT WAKULLA SPRINGS

From the surface, Wakulla Springs looks like a placid natural swimming pool, and the casual visitor would hardly suspect that startling discoveries have been made beneath the sparkling water. Yet about thirty-five years ago, someone began to suspect that the dark hulks resting some eighty feet below the surface were not sunken logs at all. With no special diving equipment, swimmers plunged to the bottom, hastily attached ropes, and hauled the objects to the surface. The "logs" turned out to be bones of mastodons, extinct relatives of elephants that lived in North America during the glacial period.

Now part of a wildlife sanctuary managed by the National Audubon Society, the springs have recently begun to yield their secrets to scientists equipped with modern diving gear. The most important discoveries have been made in a flooded cave hidden near the bottom of the springs. The sandy floor within the cave is littered with fossil bones of mastodons, giant sloths, armadillos, and other animals. No one knows when or how they got there. Possibly the animals fell in through a sinkhole that has since been plugged, or perhaps they fell into the springs and their remains later slipped down along the sloping floor into the cave. In any case, the bones give us a fascinating insight into the wildlife of Florida during the distant past.

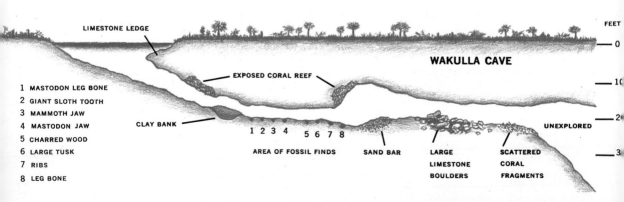

LIMESTONE LEDGE

WAKULLA CAVE

EXPOSED CORAL REEF

1 MASTODON LEG BONE
2 GIANT SLOTH TOOTH
3 MAMMOTH JAW
4 MASTODON JAW
5 CHARRED WOOD
6 LARGE TUSK
7 RIBS
8 LEG BONE

CLAY BANK

1 2 3 4   5 6 7 8

AREA OF FOSSIL FINDS

SAND BAR

LARGE
LIMESTONE
BOULDERS

SCATTERED
CORAL
FRAGMENTS

UNEXPLORED

FEET
— 0
— 1(
— 2(
— 3

The New Paris excavations

Boobytraps for the unwary but bonanzas for scientists, these deep sinkholes at New Paris, Pennsylvania, gradually filled up with animal and plant remains. By excavating debris from one of the sinkholes and recording the kinds of fossils found at various depths, researchers have reconstructed a picture of climatic changes in the area since the end of the last Ice Age.

It is the combination of plant and animal remains in undisturbed layers that scientists are always looking for, but which they rarely find. Churning floodwaters and the almost inevitable mixing that occurs when animals drop into a cave have often prevented scientists from "reading" the records of the past. One notable exception has recently been reported, however—a long-delayed follow-up of a bit of cave exploration carried out by author Charles Mohr in 1932.

Investigating a report of fossil beds in caves near New Paris, about twenty miles from Bedford, Pennsylvania, Mohr and his companion found nearly a score of shallow saucerlike depressions ten to fifteen feet across and a few sinkholes with openings the size of manhole covers. Rigging their rope to nearby trees, they lowered themselves consecutively into four of the sinkholes. Most of the way down, the sinkholes were narrow enough so that the speleologists could steady themselves against the vertical walls. But where the pits wid-

ened, the speleologists spun dizzily until they reached the bottoms—cone-shaped piles of leaves, earth, rock, and bone.

Some of the narrow fissures they explored connected with other shafts which apparently once led to the surface but which were now covered with roots and earth. Some shafts were partially filled with debris; others dropped to bare rock floors forty feet below. Evidently these had never been open to the surface; they had never served as traps for unwary travelers above. But Sinkhole No. 2, as it was later designated, had trapped several elk before the early settlers exterminated the herds. Broken antlers and bone fragments were quickly uncovered; but since Mohr and his companion were not prepared for a fossil "dig," they settled for a couple of bones as evidence of the deposit. It was not until 1946 that amateur paleontologists from Pittsburgh began a series of sporadic visits to New Paris.

Work started in earnest in 1958 when spelunkers and other volunteer bone diggers began excavating Sinkhole No. 4— Lloyd's Rockhole—200 yards from the one where Mohr found the elk bones. In five years, forty men and women, many of them of college age, took out 125 tons of earth, bones, and other fill. Supervised by Curator John Guilday of the Carnegie Museum, they sifted the debris literally by the spoonful so they would not miss the smallest tooth or bone.

The material they found, sorted, and catalogued provides the clues for an amazingly complete reconstruction of the prehistoric climate, living forms, and sequence of changes that followed the retreat of the last ice sheet, the Wisconsin, from the border of this Appalachian area.

Bones and pollen

The conclusions reached in this involved detective story are convincing for two reasons. First, there were great numbers of bones—from 2769 reptiles, amphibians, birds, and mammals—embedded in mud containing readily recognizable plant pollen and datable charcoal. Second, the bones were neatly layered. Unlike bellshaped Cherokee Cave, where fossils had been widely spread and completely churned by floodwaters, narrow shafts at New Paris confined the material as it accumulated very slowly, over thousands of years. Each new level consisted of material older than the layer above it.

Volunteers sift the debris from Sinkhole No. 4 for fossil bones and teeth of small animals. Such apparently insignificant remains often provide better clues to the composition of past populations than do the relatively scarce remains of large animals.

161

JACK PINE POLLEN

WHITE SPRUCE POLLEN

SNOWSHOE HARE SKELETON

In almost all previous cave digs, scientists had concentrated on identifying the very large mammals. Unfortunately, by their weight alone they often settled into positions that made it difficult to relate them to the age of the pollen in the datable layers. But Guilday and his assistants were masterfully thorough. By refined techniques of sorting and sieving the very dust of the deposit with infinite patience and care, they found and saved thousands of bone fragments of the smaller animals which would have gone unnoticed in earlier studies.

To everyone's surprise, the commonest remains found at New Paris were the skulls of the yellow-cheeked vole, *Microtus xanthognathus*, an eight-inch, short-tailed, shaggy brown burrowing rodent with orange-yellow fur around its muzzle. They live today in the subarctic from Hudson Bay to Alaska, but they have so seldom been found by scientists that the National Museum of Canada, with over 31,000 mammal skins in its collection, has only two. But at least 340 yellow-cheeked voles had fallen into Sinkhole No. 4, 1200 miles south of their present range.

Just as important as the animal remains was the plant pollen associated with them. Pollen grains in the humus and clay and in the scats (the droppings of mice and other plant eaters) revealed the ecological conditions under which the animals lived. And, far better than broad lakes in more open country, where wind-blown pollen can collect from great distances, the tiny sinkhole in the intermontane valley was a selective trap for the abundant pollen from the immediate area. Pollen preservation at New Paris was poor, but enough survived to give a clear record of plant succession during the thousands of years the deposit was building.

By a rare stroke of luck, the amateur paleontologists assisting Guilday found charcoal in the lower bone-bearing layer, probably from a forest fire. Guilday knew that it would be possible to have this wood dated with great precision. Like other organic material, it would contain ordinary carbon atoms, carbon 12, and its radioactive isotope, carbon 14, which decays at a known rate. The proportion of carbon 14 remaining in the charcoal would reveal just how long ago the tree died. Guilday sent some of the charred wood to Yale University's Geochronometric Laboratory, early headquarters for carbon 14 studies. The forest fire occurred about 11,300 years ago, according to the Yale analysis, soon after the end of the last glaciation, the Wisconsin.

PLATE CXV

Guilday's reconstruction of the past

All the data on the pollen grains and on the multitude of bones—skulls, jaws, femurs, and so forth—were punched on computer cards. When the cards were run through the computer, the raw data emerged as graphs. By following the changing percentages of each individual animal species from each level of the excavation and comparing them with the corresponding graph for plant pollen, Guilday was able to describe in detail how the New Paris landscape and the animals in it had slowly changed during the past 11,000 years.

The first act in Guilday's drama has a subarctic parkland setting, with widely scattered spruce and jack pine and a ground cover of lichens and tough prairie grasses. The cast includes arctic shrews, lemmings, yellow-cheeked and spruce voles, the Hudson Bay toad, and the prairie-loving thirteen-

This hand-colored lithograph of three yellow-cheeked voles is by John James Audubon, famed eighteenth-century painter of American wildlife, who worked from a Canadian specimen. Skulls of the rare yellow-cheeked voles, which presently live only in northern Canada, were the most common remains found at New Paris, Pennsylvania.

163

The animal remains and plant pollen discovered at New Paris, Pennsylvania, provide dramatic evidence that North America's climate has been warming since the retreat of the last glacier. John Guilday's thorough analysis of the debris collected from Sinkhole No. 4 revealed that the landscape around New Paris resembled a subarctic parkland (*above*) when the shaft was first opened to the surface. Gradually coniferous forests invaded the area; and as the sinkhole became filled, broadleaf trees and animals typical of more southerly temperate forests began to appear (*below*).

lined ground squirrel. The first act could just as well be set today somewhere north of Lake Winnipeg. By the middle of this act, the time of the forest fire, many of the larger animals had disappeared. Why?

"The finger points to early man," says Guilday. Man came out of the Northwest and burst on the scene with devastating impact, upsetting a situation that was already in a delicate state of imbalance resulting from the rapid recession of the ice. There was no evidence of early man in Sinkhole No. 4, but a sensational find of this kind might well be made in the next cave or sinkhole deposit.

By the second act in the "play," the scenery has shifted. White pine is present now, and broad-leaved trees: maples, birches, and oaks. New animal members also appear in the cast. The chipmunk has replaced the ground squirrel, evicted when the forest closed in, and red squirrels and northern

flying squirrels begin to travel newly established pathways through the treetops. Shorter, milder winters now permit garter snakes to live here.

By the third and final act—estimated to have taken place 8000 to 9000 years ago, on the basis of data from other paleontological sites—many more amphibians, and reptiles such as the timber rattlesnake, appear. The big northern flying squirrel is replaced by the smaller southern species, and the arctic shrew by the familiar short-tailed shrew.

Guilday's reconstruction of the past is straightforward and convincing. It appears that the New Paris Sinks were opened and began trapping animals some time after the last glaciation, about 12,000 years ago. Does this mean that the living inhabitants of other caves have evolved in the last 12,000 years? Probably it takes longer, though we cannot know for sure until fossil troglobites are found.

Isolation of cave species

The almost total absence of troglobitic species in caves in formerly glaciated areas and their regular appearance in nearby caves to the south suggest that the icing-over of caves prevented cave dwellers from getting food. Wherever caves were covered by ice, cave-adapted animals died out. No new ones have evolved since the last ice retreated, and species that survived just south of the ice have not recolonized the iced-over area.

On this basis it might be argued that troglobitic species have existed since at least before the last period of glaciation, that is, about 100,000 years. Some may even have been living in caves before the first of the four latest major glacial advances, about a million years ago. Scientists now believe, for example, that the ancestors of our commonest cave crayfish, *Orconectes pellucidus,* were isolated in caves at the beginning of the Pleistocene epoch.

For millions of years during the Pliocene epoch, these crustaceans inhabited shallow meandering streams and regularly moved in and out of caves. They were troglophiles. With the onset of the Pleistocene epoch, the land in the Appalachian region began to rise, and formerly slow-moving streams became steep, rocky, and too turbulent for crayfish. Only those populations that had already taken up residence in caves were able to survive; their descendants remained in the sluggish underground streams and gradually acquired troglobitic characteristics.

But these cave crayfish were certainly not the only animals to become isolated in caves during the Pleistocene epoch. Thomas Barr, of the University of Kentucky's Biospeleological Institute, thinks that certain species of troglobitic

In terms of a human lifetime, the evolution of cave animals seems like an ancient development; it has required at least a million years and is still in progress. Yet in terms of the entire history of the earth, their adaptation to cave life represents a relatively recent development, as does the evolution of man himself. If the entire history of the earth could be compressed into a single day, man's immediate ancestors would not appear until about forty seconds before midnight, and the evolution of modern cave animals would take place in the last twenty seconds before midnight. Some of the events leading up to these developments are shown here.

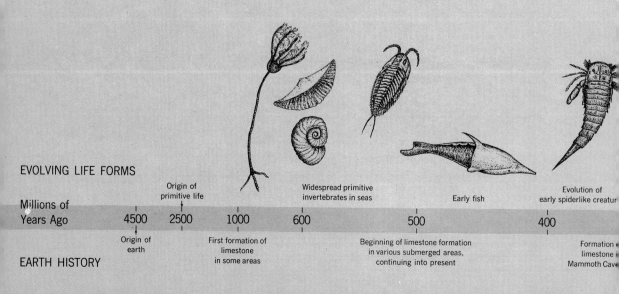

EVOLVING LIFE FORMS

| | Origin of primitive life | | Widespread primitive invertebrates in seas | | Early fish | | Evolution of early spiderlike creatur |

Millions of Years Ago   4500   2500   1000   600   500   400

| | Origin of earth | | First formation of limestone in some areas | | Beginning of limestone formation in various submerged areas, continuing into present | | Formation limestone Mammoth Cave |

EARTH HISTORY

ground beetles could have evolved after each of the four retreats of the continental ice sheets, making some of them as much as a million years old. His line of reasoning helps to explain how animals become isolated in caves.

As the New Paris story shows, during the farthest advance of the glaciers, areas just to the south of the ice evidently had a cool, damp climate like that of present-day central Canada. At the maximum advance of each successive ice sheet, there probably were spruce-fir forests in Indiana and Kentucky. The forest floor had a deep carpet of moss and lichens, a perfect habitat for ground beetles—*Carabidae*.

Barr speculates that these beetles probably lived in the damp, cool darkness under the mosses just as they do today on the mountain peaks of the Appalachians, where spruce-fir forests still survive. And among them, as among their mountaintop relatives of the present day, successive mutations made their coloring lighter, their eyes smaller, and their wings nonfunctional. They were adapted to life in environments similar to caves, and many were already troglophiles.

The retreat of each glacial ice sheet was a disaster for most of these beetles. With a warmer, dryer climate, the forest thinned out and the mosses disappeared, and with them the beetles. Indeed, the beetles were wiped out nearly everywhere except on the highest mountains where the climate was damp and cool enough for evergreen forests.

In Indiana and Kentucky, according to Barr, beetle populations living near caves or in cave entrances took shelter inside. Already adapted to darkness and a cool, stable microclimate, they survived without difficulty and have since accumulated troglobitic adaptations.

The southwestern United States also underwent striking changes with the retreat of the glaciers. Great hardwood

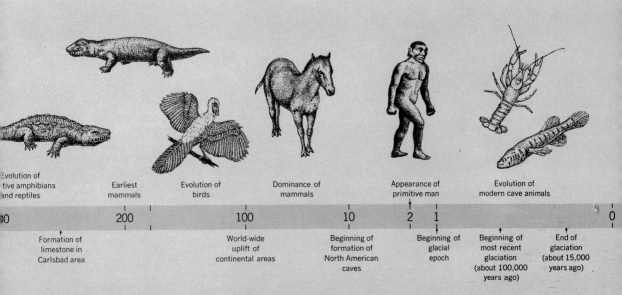

Evolution of tive amphibians and reptiles

Earliest mammals

Evolution of birds

Dominance of mammals

Appearance of primitive man

Evolution of modern cave animals

00   200   100   10   2   1   0

Formation of limestone in Carlsbad area

World-wide uplift of continental areas

Beginning of formation of North American caves

Beginning of glacial epoch

Beginning of most recent glaciation (about 100,000 years ago)

End of glaciation (about 15,000 years ago)

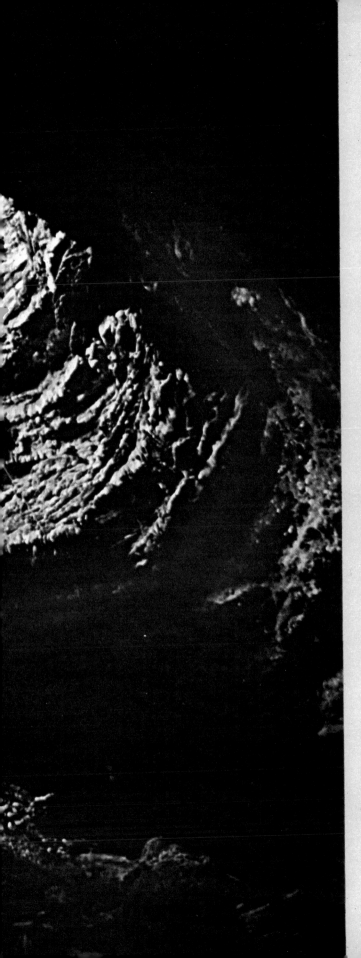

TEXAS CAVES

With about 1500 known caves and
possibly as many still to be discovered,
Texas is one of the great cave areas
in the United States. Flooded several
times by ancient seas, vast areas of the
state contain thick limestone deposits,
ideal for cave formation. Most of the
caves are in the central and western
parts of the state, but in terms of
cave-adapted animals, the richest area of
all is the Edwards Plateau in south-
central Texas. Cave dwellers here are
so abundant, in fact, that biologists
consider the Edwards Plateau one of
the most important cave regions in the
world. At least ten different kinds
of troglobitic amphipods, for example,
are known to inhabit the Edwards
Plateau, though so far only two species
have been scientifically described
and classified. Robert W. Mitchell
of Texas Technological College has
estimated that, including future
discoveries, the caves of the Edwards
Plateau may harbor as many as 200
species of true troglobites—certainly
an amazing total for any cave area,
and one to set the speleologist itching
to get out and have a look for himself.

*After lowering themselves on ropes,
two spelunkers stand on a huge
mound of breakdown debris rising
from the bottom of Devil's Sinkhole,
a 407-foot-deep vertical shaft in the
Edwards Plateau of Texas. Guano
from the sinkhole's colony of eight
million Mexican free-tailed bats
coats the rocks in the foreground.*

Found nowhere else in the world, cliff frogs are widespread
in central Texas. These inch-long amphibians so
commonly frequent cave entrances that they are sometimes
called cave frogs. Texas spelunkers often hear the little frogs
chirping from hidden crevices but seldom see them.

The inch-long assassin bug completes its entire life cycle in the entrance zones of some Texas caves. Also known as the bloodsucking cone-nose, it considers any unwary human a potential meal—and its bite is painful.

A predatory white spider prowls through the dark interior of a cavern on the Edwards Plateau. Troglobitic spiders usually stalk their prey instead of engaging in the energy-consuming practice of spinning webs.

At a length of three and a half inches, this cave-dwelling centipede is a giant of its kind; most of its kin average about two inches in length. Eyeless centipedes inhabit many Texas caves, but they are rarely abundant.

forests resembling those of Kentucky today were wiped out by a drastic drop in rainfall. They were replaced by semi-desert scrub and by cactus and mesquite. Most of the crickets that had flourished in the forests of New Mexico and Arizona died out, but a tiny remnant found their way into caves and survived.

The specialized western cave cricket known today as *Ceuthophilus longipes* is such a remnant. The cricket's pale body is at most an inch long, and its degenerate eyes have lost all function. Elongated antennae and long legs help it to obtain its food with little expenditure of energy. These troglobitic developments are not surprising. The reduced size of this cricket reflects the meager food supply in southwestern caves.

Pathways and barriers

Troglobitic cave animals are isolated in the sense that they can no longer venture out into the world of light; but they and their descendants are not necessarily confined to the caves in which they are born. Speleologists are continually discovering evidence of extended migration underground. Troglobitic species, particularly the aquatic varieties, have established stable populations in widely separated caves. Wherever caves possess navigable underground connections, there is traffic between them.

Recently Loren Wood and Robert Inger of the Chicago Natural History Museum were called upon to explain the perplexing presence of the rare cavefish *Amblyopsis rosae* in several widely separated Missouri wells. There was no visible connection between the sites nor were there any known caves in the area. Investigation satisfied them that these wells, located on two sides of a ridge, were connected by a deep water-filled channel in the limestone rock. *Amblyopsis* was living and breeding in these hidden hollows. The scientists could only guess how far the channel extended and how many inaccessible caves supporting rare forms of life existed in the region.

Like peepholes to a hidden world, sinkholes such as this one in Kentucky are striking evidence of the existence of cave systems that may extend for miles.

In areas where caves are common, deep underground channels are often encountered during drilling operations. In 1957, for example, test drilling into the bed of an upstream tributary of the Cumberland River during a survey for a damsite revealed that there were deep channels in the limestone over which the river flowed. Until recently no one suspected that troglobites used such cavities; now we know they do.

Before these river-bottom channels were filled with concrete at the damsite, fishery experts poisoned the waters with rotenone. They wanted to determine the kinds and the numbers of fish present. Rotenone acts quickly. Fish started to rise to the surface right away. Within an hour or two the fishery experts collected the fish and departed. But a day later several dozen white fish floated to the surface. They were *Typhlichthys*. Apparently the rotenone had taken a day to filter slowly down into the open interconnected channels in and below the riverbed. Evidently the fish were living in the channels or traveling between caves on the opposite sides of the river.

Troglobitic species have attained their greatest distributions where there are many caves linked together by occasional underground connections. One such region extends from southern Indiana through the Mammoth Cave area in mid-Kentucky and on through Tennessee into northern Alabama. In this Interior Lowland Plateau region there are hundreds of caves. Here certain species of troglobitic beetles have been found in caves throughout an area of 150 miles or more. And within one area, there are several different species. For example, Cumberland Caverns shelters at least six species of troglobitic beetles.

Not all the caves in the Interior Lowland Plateau are interconnected, of course. Connections become blocked and plugged; new ones develop. But over a long period of time,

Two closely related Texas amphipods, an eyed surface dweller (*top*) and a blind white cave dweller (*bottom*), are each about a quarter of an inch long. Floods carry surface amphipods over wide areas, but cave amphipods tend to be restricted to single cave systems.

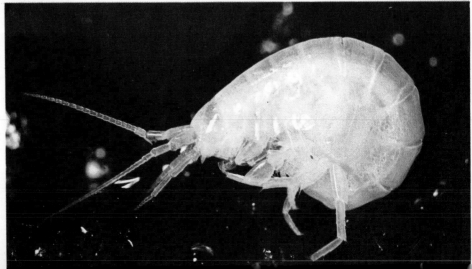

first through one passageway and later through others, movement between these caves has been possible. The troglobites have had time and opportunity to establish widely scattered colonies.

Aquatic cave animals are generally more mobile and have much wider ranges than the terrestrial cave forms. Fish especially can travel freely, with little metabolic cost, through water-filled tunnels that are impassable to land animals. The latter, when they migrate, usually crawl along the walls or floors of damp passages. Such interconnections develop only in the late maturity of a cave, and they last for a shorter time than water-filled passages. Hence land-dwelling cave animals less commonly have wide distributions.

Barriers of insoluble rock can confine a troglobitic species to an area of a few miles or even to a single cave. Recently, for example, John Holsinger, a speleologist from the University of Kentucky, studied five related but distinctly different species of cave amphipods in the mountains of Virginia and West Virginia. Each is isolated in its own series of caves in a single river valley. One valley is separated from the other by high mountain ridges that have no soluble limestone rocks. There can be no intermingling of populations. No cave amphipod can join its relations in the next valley. The only widely distributed amphipod in the area is a surface dweller. Free to be transported by floods from place to place and to move overland through springs and wet areas, it has established thriving colonies in several river valleys. It is less affected by the insoluble barriers that have confined its cave relatives to a few caves in separate river valleys.

All troglobites—but especially isolated species, such as the cave amphipods just described—are more vulnerable to disruptions in their habitats than are surface species, which may be widely scattered. They are endangered not only by sudden changes in their environment, such as pollution, mud fills caused by surface erosion, and drought, but also, as we shall see next, by impending changes in climate.

Will our warming climate affect cave animals?

Geologists say that the earth has been warming since the last glaciation. The annual temperature in North America is expected to continue to rise for about the next 25,000 years.

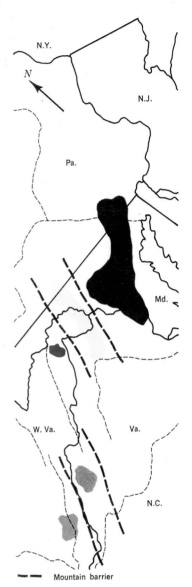

— — Mountain barrier

The ranges of five species of amphipods inhabiting caves in the central Appalachian Mountains are shown below. Four are restricted to caves in different river valleys; one (its range is shown in red) is found in a single cave and in several seeps and springs. Mountain barriers of insoluble rock have prevented the intermingling of populations.

175

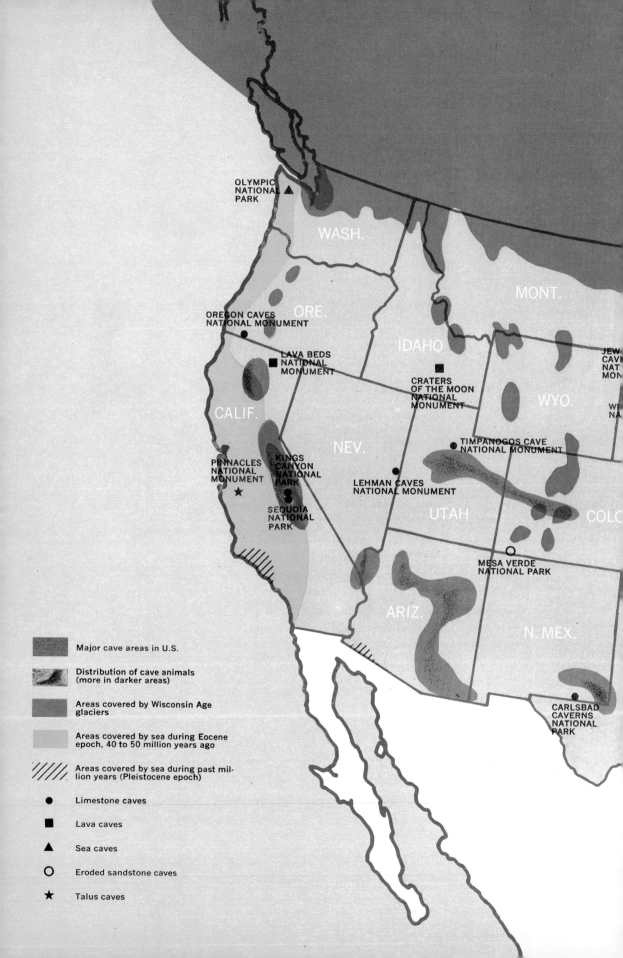

OLYMPIC
NATIONAL
PARK ▲

WASH.

MONT.

OREGON CAVES
NATIONAL MONUMENT ●

ORE.

IDAHO

JEW
CAVE
NAT
MON

LAVA BEDS
NATIONAL ■
MONUMENT

CRATERS
OF THE MOON
NATIONAL
MONUMENT ■

WYO.

WI
NA

CALIF.

NEV.

TIMPANOGOS CAVE
NATIONAL MONUMENT ●

PINNACLES
NATIONAL
MONUMENT ★

KINGS
CANYON
NATIONAL
PARK

LEHMAN CAVES
NATIONAL MONUMENT ●

UTAH

COLO

SEQUOIA
NATIONAL
PARK

MESA VERDE
NATIONAL PARK ○

ARIZ.

N. MEX.

CARLSBAD
CAVERNS
NATIONAL
PARK

Major cave areas in U.S.

Distribution of cave animals
(more in darker areas)

Areas covered by Wisconsin Age
glaciers

Areas covered by sea during Eocene
epoch, 40 to 50 million years ago

Areas covered by sea during past million years (Pleistocene epoch)

● Limestone caves

■ Lava caves

▲ Sea caves

○ Eroded sandstone caves

★ Talus caves

MAJOR CAVE AREAS IN THE UNITED STATES

No matter where a person lives in the United States, he is within one day's drive of a cave. Practically every state includes at least one cave, and a few states contain hundreds. Many of the best caves are in national parks and monuments, and others are in national forests or in state parks. Some are privately owned and are run as commercial ventures.

Scattered throughout the country are sea caves, lava tubes, sandstone caves, and talus caves lying beneath great boulders, but most common of all are limestone caves. Since large portions of North America were at one time or another submerged beneath ancient seas, thick limestone deposits are fairly common. Two relatively recent invasions by the sea are indicated on the map, but earlier ones were even more extensive.

Not every cave is inhabited by permanent cave dwellers. Areas that were covered by recent glaciers have little or no cave life; the troglobites were wiped out by the advancing ice sheets. In the relatively short time since, animals have not been able to reinvade caves in the northern areas, and new forms have not yet evolved to fill the gap. But to the south, cave life is widely distributed and occasionally even abundant.

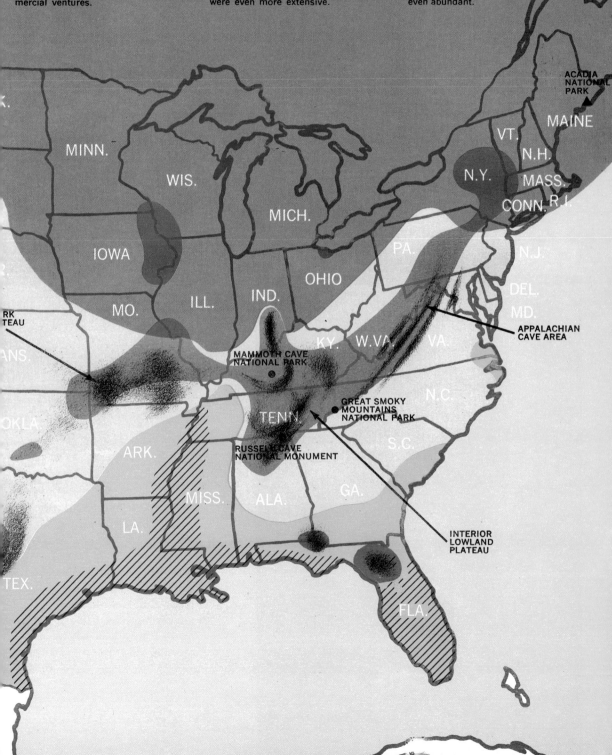

Birds and mammals typical of warmer regions will steadily extend their territory northward. By 10,000 A.D. New York will be considerably warmer than present-day New Orleans. By 20,000 A.D. Boston may have the hot, humid weather, the palmettos and pineapples, and the shrill bright parakeets of the Caribbean.

Conditions underground will change too. Caves, then as now, will maintain throughout the year the average annual temperature of the surface above them. In the future, the caves we have described in this book will have tropical climates. They will still receive water and food from the surface, but the quantity and quality of water and organic matter and the timing of their arrival are certain to be different.

You have seen that cave animals have developed economies of metabolism and behavior that compensate for the meager, irregular food supply. Low temperature slows the rate of their body processes, allowing them to live on the limited food they obtain. You can imagine that a rise of only a few degrees might speed up these processes beyond the limits of the available food. If this happened, these animals

Lush vegetation clings to the steep walls of a *sótano*, a sinkhole-type cavern common in some parts of Mexico. During the rainy season, huge waterfalls may plunge over the rims of these entrance shafts and cascade down several additional steplike drops hundreds of feet underground.

In some Mexican caves food is so abundant that surface species are able to survive very well underground. This eyed brown centipede (*opposite page*), for example, is not a troglobite, and yet it was found deep inside a cave. An active predator, it kills smaller animals by injecting poison with its clawlike front "legs."

would die. But it is likely to happen? The answer depends on the kind of climate accompanying the warming temperatures.

An abundance of food could kill cave animals

Relatively few troglobites have been found in tropical caves. Biologists on recent National Speleological Society expeditions to tropical Mexico and to the Rio Camuy Cave area of Puerto Rico noticed a great deal of organic material in the caves—so much that there is certainly no scarcity. In such caves, then, there is less of an advantage in the energy-saving behavior and adaptations that troglobites have developed.

The presence of huge populations of cockroaches, whip-tailed scorpions, spiders, and other arthropods in tropical caves caused some surprise until it was realized that the bat roosts are occupied the year around—and there are no dormant periods. Fresh guano is always available. Furthermore, during the rainy season, floods bring vast amounts of debris into the caves.

Nearly all the animals in tropical caves have eyes and pigment; most of them are troglophiles. Since climate has changed little in the last million years, few if any troglophiles have been forced to remain in caves. Even those that have been isolated long enough to lose eyes and pigment will probably turn out to have few of the specialized features associated with a meager food supply.

One of the characteristics of troglobites is their slow feeding rate. For example, if you place a troglobitic fish or cray-fish in an aquarium with a troglophilic relative, the troglobite starves. It finds the food more efficiently than the troglophile, but the troglophile rushes all over the aquarium and so fills its stomach more quickly. This fact may explain why troglobites are generally not found near cave entrances: they cannot compete with troglophiles where food is more abundant. If the warming trend continues—with a wet climate—troglophiles might eliminate troglobites completely.

In a food-rich Mexican cave, just a few sweeps of the net yield as many as thirty finger-sized Mexican blindfish. Indeed, these fish are so plentiful in certain caves that a hundred or more may be visible in just a few cubic yards of water. Three closely related Mexican cave fish and a widespread surface form are so similar that some scientists think all may be subspecies of the same animal. In any case, the fish can interbreed, and residents of some caves show all degrees of eye and pigment degeneration. A blind Mexican cave fish and its surface relative are compared on the next two pages.

A party of explorers descends to the watery depths of Rio Camuy Cave in Puerto Rico. The guano of bats that live in the cave throughout the year provides a rich and dependable food source for hordes of smaller animals.

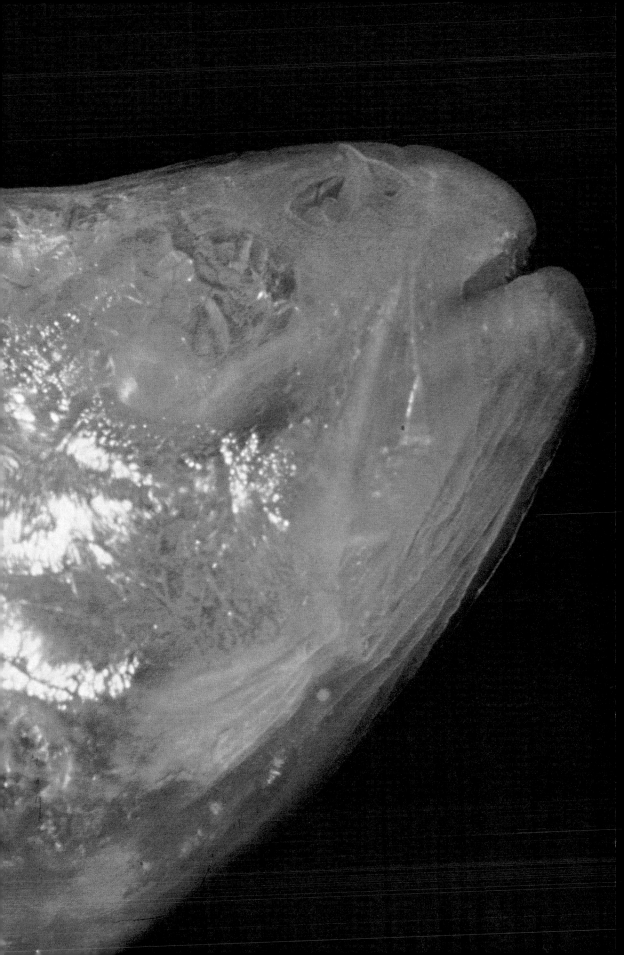

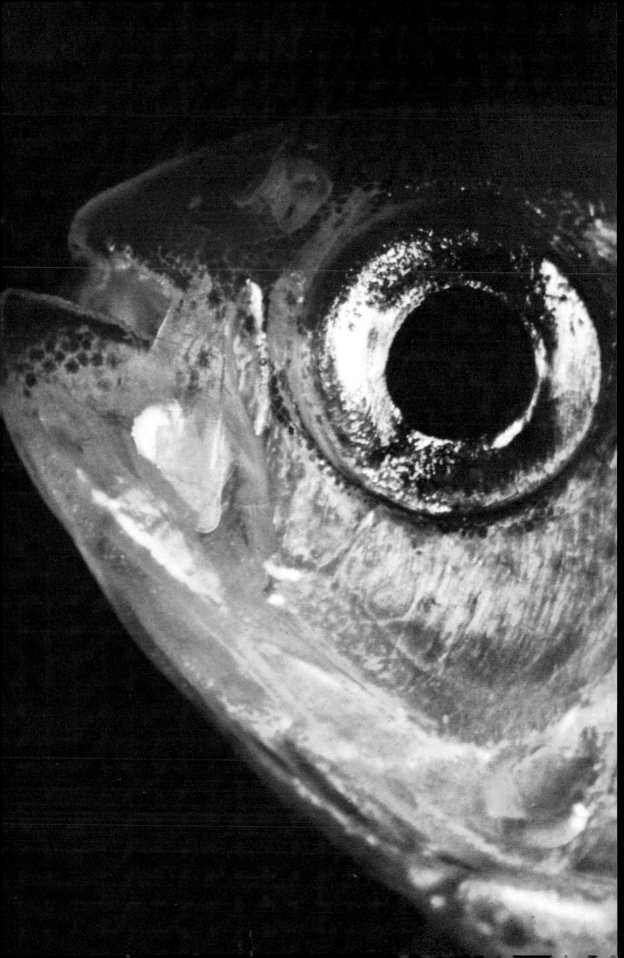

Lovers of moist, dark places, gigantic cockroaches cling to the walls of a cave in Guerrero, Mexico. Like many creatures living in tropical caves, they show none of the extreme adaptations of animals found in caves with less dependable food supplies.

But if the climate becomes warm and dry, food will be relatively scarce, and the troglobites, with their low metabolic rates, will outcompete any troglophiles that try to invade the deeper recesses of their caves.

Survivors from the sea

The few specialized tropical troglobites that are known to science are of two kinds. Either they come from caves in semiarid regions where food is scarce or they are relicts—that is, survivors—of salt-water species that have gradually invaded subterranean fresh waters.

Apparently the ancestors of these animals were not able to compete successfully with other marine animals. Only those that found their way into caves and were preadapted to life there survived. They had no competitors in the caves. As the seas receded, species isolated in this way gradually became accustomed to fresh water and accumulated troglobitic adaptations.

Marine relicts have been found in both tropical and temperate caves along former seacoasts. Some have been land-locked—separated from the sea—since pre-Pleistocene times. They have been in these caves through several cycles of Ice

Age cooling and warming, and they have undoubtedly tolerated quite extreme, though gradual, changes.

It appears, therefore, that troglobites may be less affected by heat than was long believed. In recent·experiments, blind beetles have been kept at temperatures ranging from freezing to nearly 80° Fahrenheit. They survived, and so have several species of troglobitic fish. Some of these beetles and fish have even been able to breed at temperatures 10° to 15° above or below their normal cave environment. Thus there is a good possibility that higher cave temperatures will not kill off our North American troglobites. In fact, the warming trend could produce more efficient troglobites; for if the climate becomes semiarid and cave food supplies dwindle, cave animals that acquire lower metabolic rates and consume less food will have a better chance of surviving.

But what forms could new metabolic economies take? Perhaps we can get some hints from a bizarre habitat—the deep sea.

This inch-long troglobitic isopod is a marine relict inhabiting Mexican caves along a former seacoast. It evolved from ocean-dwelling ancestors that invaded fresh-water caves when the ancient coastline retreated.

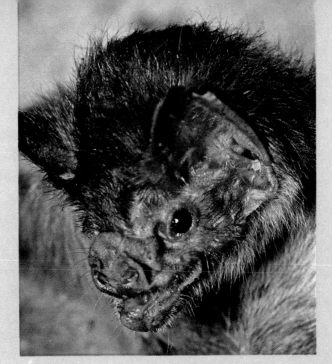

*A curiously pushed-in nose
permits the blood-drinking
vampire bat to nuzzle close enough
to its victims to slash their skin
with its slightly protruding,
razor-sharp front teeth.*

VAMPIRE BATS

Most of the fantastic legends about vampire bats are false,
but the strangest one of all is true: vampire bats drink blood.
Blood, in fact, is the only food of these uniquely specialized
inhabitants of Central and South American caves.
Incredibly lightweight despite a fifteen-inch wingspread,
a hungry vampire bat settles so gently on a cow,
goat, man, or wild beast that the sleeping victim
seldom wakens. It walks lightly in search of an exposed
spot. Then, with scalpel-like front teeth the bat shaves a
thin slice from the animal's skin and drinks the blood that
oozes from the wound. A chemical in its saliva retards
clotting of the blood while the bat takes its fill, usually
an ounce or so per night, or half its body weight. Loss
of blood seldom harms the victim, but infected bats
do spread rabies, a fact that only darkens an already
tarnished reputation.

*Hanging singly or in small clusters
by day, vampire bats emerge
at night from their roosts in
tropical caves and search for
warm-blooded prey.*

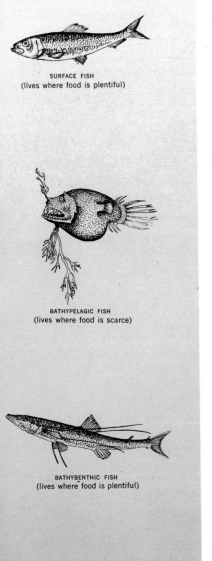

SURFACE FISH
(lives where food is plentiful)

BATHYPELAGIC FISH
(lives where food is scarce)

BATHYBENTHIC FISH
(lives where food is plentiful)

In the ocean, as in caves, certain animals are adapted to food scarcity. Food is plentiful at the ocean's surface, where microscopic plants flourish, and at the bottom, where dead bodies accumulate. In the depths above the bottom (the bathypelagic zone), however, food drifts by occasionally but does not accumulate. Fish inhabiting this zone resemble cave dwellers in their adaptations for conserving energy.

The deep sea has some cavelike features

Deep-sea fish live in a habitat much like caves, with no night and day, with a low and constant temperature, and with very little food supply. The truly deep sea, the *bathypelagic* zone, begins at a depth of about 3000 feet. At, or just below, the surface, food is produced by vast numbers of microscopic algae and other plants. However, most of it is consumed in the twilight zone which occupies the top 1000 feet. Only dead bodies fall below this level, and even this "rain" of dead organisms is mostly decomposed by bacteria and various animals before it reaches 3000 feet. Below this point food material is thinly spread. Most of it settles to the bottom, where it accumulates in considerable amounts. Even at depths of over a mile, organisms on the bottom are more varied and numerous than in the bathypelagic zone.

If you compare bathypelagic fish with those living on the deep sea floor—the *bathybenthic* fish—you will appreciate the fact that metabolic economies result from food scarcity, not from extreme depth and pressure. Obviously, the fish of the deep sea floor are exposed to greater pressures than the bathypelagic species, yet scientists have not observed any extreme metabolic economies. Why? The answer is food: on the deep sea floor there is plenty. But food does not accumulate in the open waters of the bathypelagic zone, and it is here that scientists have found organisms with the most striking adaptations to scarcity of food.

Most of the bizarre fish that live in this darkness are predators. They eat anything they can swallow, even if it is considerably bigger than they are. Everything eats everything else, and cannibalism is common. Some fish have luminescent lures hanging in front of their mouths; when they jiggle these lures, prey are attracted within reach.

Here you have the answer to an interesting question: Since deep-sea fish live in the dark, why haven't they lost their eyes as cave fish have? Eyesight is still useful in the deep sea because so many organisms produce light—luminescence. Food scarcity is the governing force. But even with eyes that help them locate their prey and with fantastic adaptations for attracting and eating them, bathypelagic fish must go a long time between meals.

Many things about the structure of bathypelagic fish suggest that they must have extremely low metabolic rates. Their gills have almost disappeared. Many species have only

188

a single pair of kidney tubules to filter wastes from the blood stream. Almost all species have lost their buoyancy control organ, the metabolically expensive swim bladder. They maintain neutral buoyancy by having reduced scales, weak skeletons, and a good deal of fatty tissue between their loosely packed muscles.

These are some of the ways of saving energy that deep-sea fish have evolved. Can we expect cave fish to evolve similar economies? There is some evidence to indicate that we can. Two different species of blind catfish have been found in deep artesian wells near San Antonio, Texas. Both species have well-developed lateral line systems and large *barbels* (whiskerlike taste organs around the mouth); and like bathypelagic fish, both have unusual amounts of fatty tissue and lack swim bladders. Unlike bathypelagic fish, however, these catfish are blind and white. They have no need of eyes in the total darkness of an artesian-well system—where there is no luminescence—and their lack of pigment is not a disadvantage.

Probably in the distant future some cave dwellers will develop metabolic economies similar to those of the deep-sea fish, but there is no way to foresee exactly what they will be. Future generations of scientists will have a fascinating subject for observation and research, provided, of course, that cave animals are not exterminated.

A menace to cave life

We dare not ignore an immediate, present menace to cave life. It is all the more disturbing because it could be avoided. Since men first discovered the troglobites, they have treated them as curiosities rather than rare and fragile products of evolution. Some species might easily be exterminated. The principles of conservation must be applied underground as well as on the surface. The National Speleological Society's

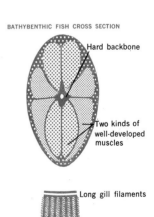

SURFACE FISH CROSS SECTION

Hard backbone

Two kinds of well-developed muscles

Long gill filaments

BATHYPELAGIC FISH CROSS SECTION

Soft backbone

Fatty tissue

One kind of poorly developed muscle

Short gill filaments

BATHYBENTHIC FISH CROSS SECTION

Hard backbone

Two kinds of well-developed muscles

Long gill filaments

Food scarcity, not depth or pressure, is responsible for the energy-conserving adaptations of bathypelagic fish. There are only minor differences between the gill and body structures of surface fish and those of bottom-dwelling (bathybenthic) fish, but the sluggish fish of the food-poor middle region show conspicuous adaptations. Their loosely packed muscle fibers are specialized for slow movement. By increasing buoyancy, soft bones and fatty tissue compensate for the loss of the energy-consuming swim bladder. Short gill filaments for extracting oxygen from water also reflect a sluggish way of life.

urgent plea "Take nothing but pictures, leave nothing but footprints" deserves to be heeded.

Pollution is particularly serious in limestone areas because of underground connections between caves. Even slight contamination is a serious health hazard and a threat to all the aquatic cave dwellers. It may not kill all the troglobites, but it could easily upset the cave community's delicate balance.

There is reason for concern. In some places, short-sighted or uninformed community officials have permitted contamination of caves. Thriving populations of cave animals have been blotted out. Prior to World War II, for instance, thousands of visitors took the subterranean boat trip in one of the Kentucky cave systems. They admired the splendid formations and looked in astonishment at the blindfish and crayfish. Now everything is changed. When tourism declined during the war years, residents of a nearby town began to use the underground river as a sewer. They piped raw sewage and industrial wastes into rock crevices that led into the underground stream. By 1946 the cave had such a nauseating stench that it could not be visited.

In 1963, when its drinking water was threatened by this underground sewer, the town finally acted. A modern sewage-treatment system was constructed, and the danger of an epidemic was averted. Now the cave will slowly return to an unpolluted condition. But it is doubtful that any of the troglobites have survived. This kind of pollution will be an increasing problem in the future.

Caves must be protected

When the population explosion reaches serious proportions in North America, public lands will be almost the only sanctuaries for natural communities. Caves will not be exceptions, particularly since spelunking is becoming a popular recreational activity.

Already national parks and monuments, national-forest primitive areas and national wildlife refuges, state parks and

Far too many people consider a sinkhole nothing more than a convenient garbage pit. Even more dangerous to the delicately balanced cave community than refuse dumping is the reckless contamination of underground streams with sewage and other pollutants.

Modern explorers reenact a scene of nearly 3000 years ago when Indians used flaming torches of cane to light their way in Mammoth Cave. Charred remains of the torches, soot deposits, and other evidence prove that the Indians traveled through many passages in search of minerals.

MAMMOTH CAVE NATIONAL PARK

Ever since its discovery by colonists in 1798, Mammoth Cave in
southwestern Kentucky has cast its spell over cave enthusiasts. Now
a part of the National Park System, the cave attracts nearly a
million visitors every year. Majestic passageways, hundreds of
intricate and beautiful formations, plus underground waterways
inhabited by blindfish, crayfish, and other creatures make the cave
a delight to both biologist and sightseer.

White men were not the first to view the wonders of Mammoth Cave,
for we know that prehistoric Indians also ventured deep into the
cavern in search of gypsum and other minerals. The Indians lighted
their way with cane torches—bundles of plant stems that burn for
hours—which left thick deposits of soot on ceilings in many passages.
Radiocarbon dating methods indicate that the charred remains of
torches found in the cave are about 3000 years old.

*In 1935 explorers discovered the
mummified body of an ancient
Indian crushed beneath a boulder
two and a half miles from the
entrance to Mammoth Cave. The
man probably was killed while
gathering minerals.*

Dwarfed by a canyonlike passage, a visitor (right) admires the area known as Hovey's Cathedral. The parallel banding of the rocks resulted from differences in the solubility of various beds of limestone deposited millions of years ago beneath an ancient sea. Interpretive exhibits in the park's Visitor Center provide a graphic account of the cave's geologic history.

With special permission from National Park Service officials, a biologist (left) nets a sample of cave creatures from Mammoth Cave's subterranean Crystal Lake. The Park Service protects dozens of important caves and their irreplaceable wildlife so that scientists of the future will be able to study undisturbed natural communities.

Broadway (below) is one of Mammoth Cave's several immense avenues, hollowed out by the dissolving action of acid-charged waters when the water table stood at a higher level. A combination of gypsum and soot from the torches of ancient Indians creates the mottling on the ceiling.

forests, and game reserves, as well as other public lands, encompass and protect the major remaining natural habitats and the living communities within them. Moreover, government conservation efforts are supplemented by some excellent private programs for protection of natural areas. Other such areas not now receiving protection should be acquired without delay. No parts of the unique Florida or Texas cave areas have yet been set aside; they should be.

We must expand the National Park System and other land-protection programs while unspoiled natural areas are still available. Mammoth Cave National Park, for example, includes no part of the sinkhole area and sinking streams that partially feed the caves in the park. They are part of the natural system, and they should be preserved and protected.

But the solution to the conservation problem is more complex than merely obtaining more land. It is easy to see that we must put aside land for recreation—for campgrounds and swimming areas—and that we must provide facilities for interpretation—nature trails, museums, displays and slide talks, and guided tours. Perhaps less obvious, but important nonetheless, is the need for natural areas that will remain as they were before man's appearance, providing guidelines for preservation and restoration, and serving as living museums for comparison with disturbed or managed areas. Their scientific

Though none of it is included in the national park, the gently rolling sinkhole plain south of Mammoth Cave *(left)* is of vital importance to the cave system. By placing dyes and other markers in sinkhole streams, scientists have shown that water from the plain flows through caves in the park, shaping caverns and replenishing their food supply, before draining into the Green River *(right)*.

values as well as their recreational values are tremendously important.

We cannot explain the intricacies of natural communities unless we understand them. And we cannot understand them unless they are available in an undisturbed state, protected from man's influence. We must assign different values to various parts of our public lands. Actually, they must be zoned for their best public use. Some areas should be out-of-bounds except to trail riders or to hikers. And perhaps some areas should be off-limits to all except qualified scientists who seek the knowledge necessary to explain the dynamics of the natural community.

The surface areas of Mammoth Cave National Park provide an example of a region where zoning is needed. Mammoth Cave Ridge represents a recreation area that is zoned for heavy public use. Much of Mammoth Cave, too, is heavily used. But portions of both the surface and the cave within the park should remain undisturbed.

Without such specially protected areas at Mammoth and elsewhere, future generations will have an incomplete picture of some of the world's largest and most important cave communities, and our discussions about the future of cave adaptations will be little more than idle speculation. The animals will not be around for future generations to examine, and our descendants will not be able to refute or uphold our theories and predictions about the life of the caves.

Scenes of awesome beauty await those who venture into the mysterious realms beneath the surface of the earth.

Appendix

A Guide to Scientific Names

CAVE CRICKET
(Ceuthophilus stygius)

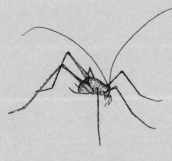

CAVE CRICKET
(Hadenoecus subterraneus)

Scientific names are an international language. Mention *Myotis sodalis* and biologists throughout the world will know you are referring to a certain kind of bat that lives in the eastern United States and is known for its habit of hibernating in great masses. But if you begin to talk about the Indiana bat, the social bat, or the cluster bat, confusion is likely to result. Your listener may have no idea that all these names refer to just one creature, *Myotis sodalis.*

If you begin to talk about cave crickets, on the other hand, a listener may think you are referring to the species known as *Ceuthophilus stygius.* In reality you may be discussing an entirely different species, *Hadenoecus subterraneus.* Both animals have the same common name. Still other animals are so unfamiliar to most people that they have no common names at all.

The only way to avoid misunderstanding is to use scientific names. They are a simple and effective means for accurate communication. Every single one of the hundreds of thousands of different kinds of plants and animals in the world has its own distinctive scientific name. By using these names, we eliminate the possibility of confusion.

If the names sometimes seem strange, this is simply because they are based on a foreign language—Latin, or latinized forms of words from Greek and other tongues. Yet the names are not as difficult as they appear. Many scientific names are used in everyday conversation. Chinchilla, gorilla, python, octopus, zinnia, petunia, and many other familiar words are parts of the scientific names of well-known plants and animals.

The scientific name itself consists of two words. The first is capitalized, the second is not. Use of this binomial ("two-name") system of classification began with Carl Linnaeus, an eighteenth-century naturalist. Before his time, scientific names often included strings of adjectives that ran on for several lines. By limiting each name to just two words, one for genus, the other for species, the business of naming was vastly simplified.

But what about the two bats *Myotis sodalis* and *Myotis lucifugus?* Why is the word *Myotis* used for both? This is because scientific classification is more than a filing system. It is also an attempt to describe evolutionary relationships. *Myotis sodalis* and *Myotis lucifugus* are two different species of bats. Yet they have certain characteristics in common that distinguish them from other kinds of bats. Both therefore are placed in the same genus,

Myotis, since they presumably are more closely related to each other than to bats belonging to the genus *Pipistrellus*, the genus *Tadarida*, or any other genus. (Note that we sometimes refer to an animal only by its genus name. By calling a bat simply *Myotis*, we mean that it is one of the several species of *Myotis* bats without specifying exactly which one.)

The next higher grouping of organisms is the family. *Myotis*, *Pipistrellus*, and several other genera (the plural form of *genus*) are lumped in the family Vespertilionidae, the so-called evening bats. The genus *Tadarida*, on the other hand, belongs to the family Molossidae, the free-tailed bats. Again, genera that are grouped in the same family are thought to be closely related.

All families of bats in turn belong to the order Chiroptera (flying mammals), a subdivision of the still larger class Mammalia (mammals). Other orders of mammals include the order Rodentia (rodents), the order Carnivora (flesh-eating mammals), and so on. Members of each order share characteristics that set them apart from all other orders.

Classes in turn are grouped into phyla. Thus, mammals, fish, amphibians, reptiles, birds, and certain other animals all belong to the phylum Chordata, animals that possess notochords (primitive backbones) at some stage in their life histories. All the phyla of animals—chordates, arthropods, protozoans, and so on—are placed finally in the overall animal kingdom. The other great kingdom of living things, the plants, represents an entirely separate line of evolution.

Thus when we refer to the bat *Myotis sodalis*, we are in effect summarizing all the following information about it:

Kingdom: Animalia (animals)
  Phylum: Chordata (animals with notochords)
    Class: Mammalia (mammals)
    Order: Chiroptera (flying mammals, or bats)
  Family: Vespertilionidae (evening bats)
  Genus: *Myotis* (mouse-eared bats)
Species: *Myotis sodalis* (social or Indiana bat)

Relationships of the slimy salamander, *Plethodon glutinosus*, in contrast, are as follows:

Kingdom: Animalia (animals)
  Phylum: Chordata (animals with notochords)
    Class: Amphibia (amphibians)
    Order: Caudata (tailed amphibians or salamanders)
  Family: Plethodontidae (lungless salamanders)
  Genus: *Plethodon* (woodland salamanders)
Species: *Plethodon glutinosus* (slimy salamander)

PIPISTREL
(Pipistrellus subflavus)

SOCIAL BAT
(Myotis sodalis)

Animals referred to by scientific names in this book are listed below, along with a generally accepted common name wherever possible. Although there is no "right" or "wrong" way to pronounce the names, one common pronunciation for each is indicated.

PHYLUM PLATYHELMINTHES (plah-tee-hel-MIN-thees): flatworms
    Sphalloplana percaeca (sfal-o-PLAN-ah per-SEE-ka)

PHYLUM MOLLUSCA (mol-LUS-ka): molluscs
    Oxychilus (ox-ee-KILL-us): snail

PHYLUM ARTHROPODA (ar-THROP-o-dah): joint-legged animals

Crayfish
    Cambarus bartoni (cam-BAH-rus BAR-ton-i)
    Orconectes pellucidus (or-ko-NECK-tees pel-LEW-sid-us)
    Procambarus pallidus (pro-CAM-bar-us PAL-ih-dus)
    Troglocambarus maclanei (tro-glow-CAM-bar-us
      mac-LANE-i)

Insects
  Order Orthoptera (or-THOP-ter-ah): grasshopperlike insects
    Ceuthophilus longipes (sue-THOF-ill-us LAWN-gee-pes):
      western cave cricket
    Ceuthophilus stygius (sue-THOF-ill-us STIH-gee-us):
      cave cricket
    Hadenoecus subterraneus (had-DEE-nah-cus
      sub-ter-RAIN-ee-us): common cave cricket
  Order Coleoptera (coal-ee-OP-ter-ah): beetles
    Family Carabidae (car-RAB-ih-dee): ground beetles
      Neaphenops (nee-AH-fen-ops): cave beetle
      Rhadine subterranea (rah-DIN-ee sub-ter-RAIN-ee-ah)
    Family Catopidae (kah-TOE-pih-dee): dung or scavenger
    beetles
      Bathysciola schiodtei (bath-ee-see-O-la shy-O-tee-i)
      Speonomus longicornis (spee-o-NO-mus lawn-gee-CORE-nis)
    Family Dermestidae (der-MESS-tih-dee): dermestid beetles

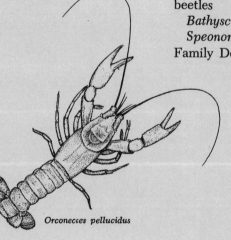

Orconectes pellucidus

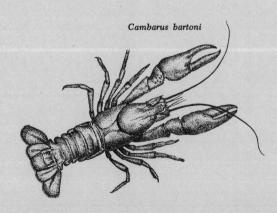

Cambarus bartoni

PHYLUM CHORDATA (core-DAY-tah): animals with notochords

Fish

    Amblyopsis rosae (am-blee-OP-sis ROSE-i): Rosa's blindfish

    Amblyopsis spelaea (am-blee-OP-sis spee-LAY-ah):
      big blindfish

    Chologaster agassizi (KO-lo-gas-ter ag-ah-SEES-i): springfish

    Typhlichthys subterraneus (tiff-LICK-these
      sub-ter-RAIN-ee-us): small blindfish

Amphibians

    Eurycea rathbuni (you-ree-SEE-ah RATH-bun-i): Texas blind
      salamander; formerly called *Typhlomolge rathbuni*
      (tiff-lo-MOLE-ghee RATH-bun-i)

    Plethodon glutinosus (PLETH-o-don glue-tin-O-sus): slimy
      salamander

    Typhlotriton spelaeus (tiff-lo-TRY-ton spee-LAY-us): Ozark
      blind salamander

Birds

    Steatornis caripensis (stay-ah-TOR-nis cah-rih-PEN-sis):
      oilbird

Mammals

  Order Chiroptera (kir-OP-ter-ah): bats

    Eptesicus fuscus (ep-TESS-ih-cus FUSS-cuss): big brown bat

    Myotis grisescens (my-O-tis grih-SESS-ens): gray bat

    Myotis lucifugus (my-O-tis lew-SIF-few-gus): little brown
      bat

    Myotis sodalis (my-O-tis so-DAL-is): social bat

    Pipistrellus subflavus (pip-ih-STREL-us sub-FLAY-vus):
      pipistrel

    Tadarida brasiliensis (ta-DAH-rih-da brah-zil-YEN-sis):
      free-tailed bat

  Order Rodentia (row-DEN-sha): rodents

    Microtus xanthognathus (my-CROW-tus zan-tho-NAY-thus):
      yellow-cheeked vole

    Neotoma (nee-o-TOE-mah): pack rat

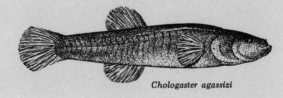

Chologaster agassizi

Amblyopsis spelaea

Caves in the National Park System

From Maine to California, North and South, East and West, the United States mainland contains a unique assortment of caves—limestone caves, of greatest importance and number, as well as lava caves, sea caves, and talus caves. A cave is a fragile resource, and its biological and geological value may be easily destroyed. There are many areas, especially in Texas and Florida, where the cave systems are not protected. Fortunately, however, a number of outstanding caves exist within units of the National Park System, and their unique living communities and rock formations are carefully preserved.

In many of the larger caves in the National Park System regular tours are conducted by trained naturalists. Elevators, graded walkways, and artificial lighting make a cave tour an exciting experience for even the least hardy of visitors. Interpretive exhibits supplement the tours by tracing the geological and archeological history of the caves and by giving some attention to their animal life. Fine vacation facilities in or near the parks include lodges, cabins, camping and picnic areas, hiking trails, and sometimes equipment for boating and fishing.

Even in caves open to the public, however, Park Service policy restricts visitors to established trails. Only qualified speleologists conducting approved research projects are allowed to visit the labyrinths of passages and chambers situated beyond the established tourist routes. These restrictions are necessary because each cave represents a unique natural laboratory where scientists can study a specific cave environment and its inhabitants under unspoiled conditions.

Acadia National Park (Maine)

An island park with a rare blend: forested mountains next to the surging sea. Here, on the eastern shore of Mount Desert Island, the pounding surf has carved Anemone Cave eighty-two feet into seemingly indestructible rock. At low tide vacationers roam through the dim interior; in the shaded tide pools they find the sea anemones that give the cave its name, as well as barnacles, sea stars, snails, and many other strange and colorful creatures.

Carlsbad Caverns National Park (New Mexico)

One of the world's largest and most spectacular caverns. Splendid formations fill its huge passages and the most tremendous rooms found in any caves. Prehistoric Indians left wall paintings inside the natural entrance, but the first settlers to become interested in the cave were bat guano miners. They eventually removed tens

SEA STAR
Anemone Cave

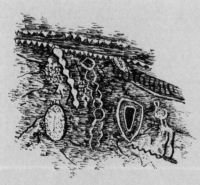

INDIAN WALL PAINTINGS
Carlsbad Caverns

206

of thousands of tons of fertilizer from an isolated section of the cave. The beauty of the caverns was not really discovered until 1924, when scientist Willis T. Lee conducted surveys that led to the inclusion of the cave and surrounding areas into the National Park System. Visitors now travel down to the 800-foot level and back by elevator. A fine Visitor Center provides information about the cave, its origin, and the animals that escape the rigorous desert environment by seeking shelter in the cave. See pages 71 to 77.

Craters of the Moon National Monument (Idaho)
An area in southern Idaho suggesting a fantastic moonscape. A great variety of volcanic phenomena can be seen: craters, volcanic cones, lava flows, and lava caves and tunnels, some of them containing permanent ice. The largest cave has about 4400 feet of passages. Cave tours are self-guided.

Jewel Cave National Monument (South Dakota)
Like nearby Wind Cave and several privately operated caves in the Black Hills, Jewel Cave has notable mineralogical features. Here crystals of dogtooth calcite stud many of the walls and ceilings. Ranging from light brown to dark chocolate, the pyramidal crystals glisten like jewels in the light of lanterns carried by visitors. A colony of lump-nosed bats winters in the cave, and pack rats are seen occasionally. Recent explorations and mapping by speleologists have shown that the cavern is far more extensive than anyone had imagined. It has been suggested that this cave was partly formed under the high pressures of artesian waters, an unusual mechanism for limestone cave formation.

Lava Beds National Monument (California)
A spectacular assortment of volcanic phenomena aboveground: lava flows, cinder cones, and spatter cones. Here is the world's greatest concentration of lava tubes or caves, at least 300 in number; some are over a mile long, and some are on four or five levels. The landscape is scarred in places by huge trenches as much as a hundred feet deep where cave ceilings have collapsed, and the trenches are spanned in many places by impressive natural bridges. Despite an arid climate, mosses, ferns, and colorful lichens flourish around and inside some of the moist cave entrances. Pack rats and a few bats constitute the bulk of the animal population. Ice can be found in the lowest levels of several of the caves. Tours are self-guided.

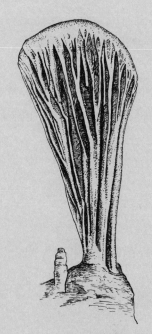

THE PARACHUTE
Lehman Caves

Lehman Caves National Monument (Nevada)
A small gem of a cavern in the side of Wheeler Peak (13,063 feet), one of the highest in the Great Basin. The cave itself is a maze of rooms and passages dissolved in marble, a form of limestone that has been altered by intense heat and pressure. A

wealth of stalactites, stalagmites, bacon rind, flowstone, rimstone dams, and many other speleothems decorate the rooms. Of special interest are the relatively uncommon formations known as shields or palettes, thin disks of calcite that angle from walls and floors. In addition to regular tours, a unique spelunkers' tour of the undeveloped parts of the cave is offered by the park naturalists.

Mammoth Cave National Park (Kentucky)

Probably the world's most famous cave, and part of one of the most extensive cave systems ever explored. Miles of giant canyon-like passageways, underground rivers, and other features of great geological and historical interest are revealed to visitors on several guided tours of varying length. Besides the attraction of blindfish, crayfish, crickets, and other cave-dwelling animals, the cave contains interesting mementos of past human activity: Indian remains, equipment where saltpeter was mined during the War of 1812, and two stone huts that served as the world's first but least successful tuberculosis hospital. Campgrounds, self-guiding trails, and the Visitor Center from which the tours start are located on the surface amid pleasantly rolling woodlands along the Green River. See pages 192 to 198.

PACIFIC GIANT SALAMANDER
Oregon Caves

Mesa Verde National Park (Colorado)

A famous showplace of prehistoric Indian culture. Spectacular cliff dwellings built thousands of years ago in almost inaccessible shelter caves are still in good condition. The largest of the dwellings is Cliff Palace, with more than 200 rooms perched on the lip of a single cavity in the sandstone cliff. Other prehistoric rock-house shelters are found throughout the Southwest, with notable examples preserved at Canyon de Chelly and Navajo National Monuments in Arizona. Another interesting dwelling, Montezuma Castle National Monument (Arizona), occupies a rock shelter eroded in a limestone cliff. Most of the shelters were abandoned about 600 years ago, when a prolonged drought cut off all water supplies.

Oregon Caves National Monument (Oregon)

A modest but attractive cave situated high on the slopes of the Siskiyou Mountains. Its passages and galleries were dissolved in marble and contain large stalagmites in abundance, as well as stalactites, columns, flowstone, and dogtooth spar. Although the cave has no permanent residents, the foot-long Pacific giant salamander is seen occasionally, as well as pack rats and eight species of bats, including the western lump-nosed bat.

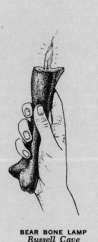

BEAR BONE LAMP
Russell Cave

Pinnacles National Monument (California)

A spectacularly rugged landscape that resulted from ancient volcanic activity. Several steep-walled canyons are accented by towering rock spires 500 to 1000 feet high. The "caves" were formed long ago when great chunks of lava rock were dislodged and became wedged between narrow canyon walls. The area known as the Big Room is formed by a hundred-foot-long lava block estimated to weigh 60,000 tons.

Russell Cave National Monument (Alabama)
Preserved in sediments on the floor of this recently studied cave is a continuous record of occupation by prehistoric Indians for 8000 years. By excavating debris from a trench over twenty feet deep, archeologists have recovered more than two and a half tons of tools, weapons, animal bones, charcoal from campfires, and even the buried skeletons of a few of the Indians. The mass of artifacts records a slowly changing mode of life since ancient men first sought shelter here about 6500 to 7000 B.C. Interpretive exhibits in the cave and Visitor Center summarize this well-documented story.

SPEAR POINTS
Russell Cave

Sequoia and Kings Canyon National Parks (California)
A number of limestone and marble caves, not so well known as the giant sequoia trees and the spectacular mountain and canyon scenery of these adjacent Sierra parks. Most notable is Crystal Cave in Sequoia National Park. Its broad entrance on a wooded mountainside overlooks a sparkling stream. Park naturalists conduct tours through the cave, which is dissolved in white marble.

Timpanogos Cave National Monument (Utah)
Three limestone caves on the slope of a snow-capped peak in the Wasatch Mountains, now connected by man-made tunnels. Though small, the chambers are noted for their colorful stalactites, stalagmites, and striped draperies reflected in pools of water. The walls are encrusted with glistening pink and white crystals and delicate twisting helictites. Animal life is sparse except for springtails, cave crickets, and a few lump-nosed and other bats.

BOXWORK
Wind Cave

Wind Cave National Park (South Dakota)
On the surface, rolling grasslands populated by bison and pronghorns and a notable prairie-dog town; underground, a large limestone cavern known for its unique mineral formations. The cave is named for air currents that blow alternately in and out of the entrance as a result of changes in barometric pressure. In summer conducted tours lead the visitor 300 feet below the surface, where narrow passages and small chambers are decorated with colorful calcite crystals. The cave is famous for its abundant boxwork, a formation common only in the Black Hills. The crystals formed in intersecting cracks in the limestone, then remained as a honeycomb pattern of projecting blades when part of the limestone dissolved away.

209

Commercial Caves in the United States

Throughout the United States, nearly 150 caverns have been developed by private owners as commercial caves. Many exceptional caves also have been preserved in state and local parks. Limestone caves, lava caves, and sea caves, large and small, are open to the public. Every one is unique; each one has a slightly different history and a combination of features that can be seen nowhere else in the world.

In most commercial caves, of course, formations are the principal attraction. Because of artificial lighting and the constant movement of people, the natural community of life usually is disrupted, and animals tend to be sparse along the tourist routes. Even so, the owner or guide may be willing to show you areas where cave animals continue to flourish.

The list below is only a sampling of attractions at some of the better-known public caves. Since most of them advertise along nearby highways and are marked on road maps, they are easy to find.

ALABAMA

Cathedral Cavern (*Grant*): A magnificent and spacious cave with exceptionally fine formations.

ARIZONA

Colossal Cave (*Vail*): A large dry cave in the desert foothills, used by migrant Mexican hognose and longnose bats as a nursery in summer, when they feed on the pollen of saguaro and other cactuses. Once the cave was used as a hideout by train robbers.

ARKANSAS

Diamond Caverns (*Jasper*): One of several public caves in the Ozarks, where caverns are large and generally very attractive.

CALIFORNIA

La Jolla Caves (*La Jolla*): Small but interesting sea caves with a variety of marine life.

Mercer Caverns (*Murphys*): A large and attractive cave with many delicate aragonite crystals and other formations.

Mitchell's Caverns (*Mitchell's Caverns State Park, Essex*): Two small adjoining caverns in the Mojave Desert, with large columns and draperies.

Moaning Cave (*Vallecito*): Contains a large room whose size seems doubly impressive because the visitor enters from the top and descends a spiral staircase. Many prehistoric human bones have been found here.

COLORADO

Cave of the Winds (*Manitou Springs*): Good-sized cave near the foot of Pike's Peak, especially famous for its display of very rare and delicate aragonite helictites.

FLORIDA

Florida Caverns State Park (*Marianna*): One of the few places in Florida with sizable caves that are not filled with water.

IDAHO

Minnetonka Cave (*St. Charles*): Largest limestone cavern in the area, but relatively undeveloped. Visitors must carry lanterns down stairs and trails that lead far underground.

ILLINOIS
Cave-in-Rock (*Cave-in-Rock State Park*): Modest cave in a limestone bluff overlooking the Ohio River. Its grisly history includes use by river pirates, counterfeiters, and murderers. The cave itself is disappointing, except for its fine arched entrance.

INDIANA
Bronson–Donaldson and Twin Caves (*Spring Mill State Park, Mitchell*): Boat trips are conducted through these pleasant caves populated by blindfish, blind crayfish, and other animals.

Wyandotte Cave (*Wyandotte*): A large cave with impressive corridors and rooms. There are relatively few formations, yet the Pillar of the Constitution, thirty-five feet high and seventy-five feet around, is one of the largest stalagmites anywhere. There is abundant evidence that Indians used the cave.

IOWA
Crystal Lake Cave (*Dubuque*): One of several caves in northeastern Iowa, this large limestone cavern is noted for its delicate formations.

KENTUCKY
Carter Caves (*Carter Caves State Park, Olive Hill*): This medium-sized park in extreme northeastern Kentucky includes a dozen caves, three of which are electrically lighted and have guided tours.

Cascade Caverns (*Carter Caves State Park, Olive Hill*): An interesting cave just outside Carter Caves State Park, operated commercially.

Diamond Caverns (*Park City*): The last of the early celebrated commercial caves in the Mammoth Cave group still operated privately (though a number of new caves have been opened in the last few years).

A wealth of large and colorful formations are found here.

Mammoth Onyx Cave (*Horse Cave*): A real gem of a cave, with a continuous gallery of speleothems; noted for highly interesting, informative tours.

MARYLAND
Crystal Grottoes (*Boonsboro*): A small but interesting cave, well decorated with speleothems.

MICHIGAN
Bear Cave (*Buchanan*): This tiny cave is unique in its origin: 4000 years ago a channel formed through a tufa deposit at a time when Lake Michigan was at a higher level. Fine leaf casts, tree limb molds, and miniature grottoes are features of prime geological interest.

MINNESOTA
Mystery Cave (*Spring Valley*): Part of a large limestone cave system in the southern part of the state. It contains red, brown, white, and yellow flowstone.

Niagara Cave (*Harmony*): This cave, considered the biggest in the upper Midwest, contains large canyonlike passages and some fine speleothems.

MISSOURI
Bridal Cave (*Camdenton*): One of more than a score of commercially operated caves in the state. It contains particularly fine speleothems.

Cherokee Cave (*St. Louis*): A former brewery cave, famous for the fossil bones discovered in it. It is unusual for its location beneath the streets of a large city.

Fishers Cave (*Meramec State Park, Sullivan*): One of several caves in this state park. Other commercial caverns are nearby.

Mark Twain Cave (*Hannibal*): A cave

whose labyrinth of passages will be delightfully familiar to any reader of *Tom Sawyer*. Mark Twain's own boyhood adventures in this cave inspired his fiction.

Marvel Cave (*Branson*): A fine cave where the visitor descends through an opening in the bottom of a sinkhole and enters a spectacular domed chamber from above. Side passages lead still further underground. An electric tramway brings the visitor back to the surface. Ozark blind salamanders are occasionally seen.

Meramec Caverns (*Stanton*): A large cave crammed with truly extraordinary formations. The cavern reportedly was used as a way station for escaping slaves on the Underground Railroad.

Onondaga Cave (*Leasburg*): Like many Ozark caves, this one has many exceptional formations. The Lily Pad Room, in particular, is superlative.

MONTANA
Lewis and Clark Caverns (*Lewis and Clark State Park, Whitehall*): The largest developed cavern in the state. There are many fine formations in its large rooms.

NEW HAMPSHIRE
Lost River Glacial Caverns (*N. Woodstock*): These caves under jumbled heaps of granite boulders were formed as a result of glaciation. Nearby Polar Caves, at Rumney Depot, are similar.

NEW MEXICO
Perpetual Ice Cave (*Grants*): Year-round ice formations in a lava tube high in southern mountains. This is one of the most accessible ice caves in the Southwest.

NEW YORK
Howe Caverns (*Howes Cave*): The largest limestone cave in New York and New England. There are varied formations and an underground stream which the visitor can tour by boat. Smaller public caves nearby

include Secret Caverns at Cobleskill and Knox Caverns at Altamont.

NORTH CAROLINA
Linville Caverns (*Ashford*): The only limestone cave open to the public in this region just east of the Appalachians.

OHIO
Ohio Caverns (*West Liberty*): An attractive limestone cave noted for its spectacular pure white stalactites and stalagmites. (The deposits in most caves are tinted by mineral impurities.)

OKLAHOMA
Alabaster Caverns (*Alabaster Caverns State Park, Freedom*): Fine gypsum caves in a state park. Their geological history is similar to limestone caves, except that the cavities developed in gypsum, which is even softer and more soluble than limestone.

OREGON
Lava River Caves (*Lava River Caves State Park, Bend*): An impressive lava tube about one mile long. Tours are self-guided, but lanterns are supplied.

Sea Lion Caves (*Florence*): A huge sea cave where visitors can watch and listen to a herd of barking sea lions. Sea birds nest in the cave.

PENNSYLVANIA
Penn's Cave (*Centre Hall*): A pleasant and historic cave where the visitor tours entirely by boat. Nearly a dozen modest but attractive commercial caves are scattered across the state. The best known are Woodward Cave near Woodward, Indian Echo Caverns near Hummelstown, and Crystal Cave near Kutztown.

SOUTH DAKOTA
Rushmore Cave (*Keystone*): This is the best place in the Black Hills to see stalactites and stalagmites, since they are scarce in most other caves in the area. On the

other hand, it lacks the crystals that are so well developed in nearby caves such as Sitting Bull Crystal Caverns at Rapid City and Stage Barn Crystal Caverns at Piedmont.

TENNESSEE

Bristol Caverns (*Bristol*): One of the finest displays of large formations in the region.

Cumberland Caverns (*McMinnville*): A vast and magnificent cave known since the early 1800s. Yet its true size and beauty were unsuspected until the 1950s, when ardent spelunkers ventured into its extensive series of chambers.

Lookout Mountain Caves (*Chattanooga*): Access to lower levels of this cave is by elevator. It contains the spectacular Ruby Falls in a great domepit.

The Lost Sea (*Sweetwater*): Boat trips are made on an underground lake 600 feet long, 150 feet wide, and as much as 60 feet deep. This cave was opened in 1965.

Wonder Cave (*Monteagle*): One of the well-known caves of east Tennessee, with several levels and many formations.

TEXAS

Caverns of Sonora (*Sonora*): A place of fragile, almost unbelievable beauty. Few caves offer such a great variety of rare and splendid speleothems.

Texas Longhorn Cavern (*Texas Longhorn State Park, Burnet*): An extensive cave with few dripstone formations, but the sculptured walls are studded in places with large calcite crystals. Recent fossils have been discovered here.

Wonder Cave (*San Marcos*): A small cave with few formations. But a well in the cave reaches the water of the Purgatory Creek system, which is known for its unique cave animals, including the famed Texas blind salamander.

VIRGINIA

Endless Caverns (*New Market*): Not really endless but certainly large; notable for the variety of formations, like other caves in the Shenandoah Valley.

Grand Caverns (*Grottoes*): Most unusual of the varied formations are shields, thin disks of calcite that jut from the walls. One, the Bridal Veil, is eight feet in diameter, with a graceful curtain of calcite hanging from its edge.

Luray Caverns (*Luray*): A large cave noted for its exceptional abundance of speleothems. In many places the multicolored stalactites and stalagmites are so densely crowded that they resemble curtains.

Melrose Caverns (*Harrisonburg*): Noted for its use as a barracks for an Ohio regiment during the Civil War. Its famous "registry column" bears the signatures of scores of northern soldiers.

Skyline Caverns (*Front Royal*): A beautiful cavern whose prize attraction is a unique display of anthodites, clusters of four-inch white calcite needles that cover the ceilings of several small rooms.

WEST VIRGINIA

Organ Cave (*Ronceverte*): One of three commercial caves in the state and part of an enormous cave system, though most is closed to tourists. Besides formations, the visitor can see perfectly preservd century-old equipment used by saltpeter miners.

WISCONSIN

Cave of the Mounds (*Blue Mounds*): This is the best known of four commercial caves in the state. The others are Crystal Cave, Spring Valley; Eagle Cave, Muscoda; and Kickapoo Caverns, Wauzeka.

WYOMING

Spirit Mountain Caverns (*Cody*): Formerly known as Frost Cave, this is the only limestone cavern in the state open to the public.

How to Become a Spelunker

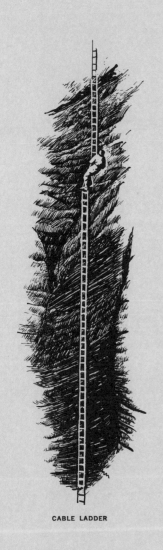

CABLE LADDER

How do you go about exploring a wild cave? What kind of equipment do you need and how do you use it? The best advice for the beginner is to proceed with caution. Start with caves that have level walk-in passages, and acquire plenty of experience before attempting to explore more difficult caves. Serious spelunking accidents are relatively rare, yet they can and do happen. There are risks, but they can be minimized by proper preparation and by recognition of the hazards involved in caving.

Begin by visiting public caves in the national parks and forests and in state parks, or commercial caves operated by private owners. Here you travel with a tour group led by a competent guide. Paths, stairways, bridges, and even elevators assure easy access and sure footing in difficult areas. Well-placed lights display the cave and its formations to the best advantage. Without investing in any special equipment, you will soon be able to decide if you really want to explore caves on your own.

If you enjoy your first taste of caves and want to learn more about them, your next step should be to join a group of experienced cavers. The nearest museum probably can put you in touch with any caving club located in your area. The best source of information, however, is the National Speleological Society, a nonprofit organization founded in 1941 for the study and preservation of caves for scientific, scenic, and recreational purposes. The society's office is located at 609 Meadow Lane, Vienna, Virginia. Its records and publications provide a national depository for information about caves, and its leaflets on safety and conservation will be provided to anyone who requests them. Local chapters of the society, known as grottoes, have been established throughout the country. The society can provide the addresses of grottoes in your area.

Rules for safe and considerate spelunking, you will discover, are more specific and rigid than those for many other forms of outdoor recreation. Yet most of them are basically matters of common sense and courtesy. The following notes will help prepare you for your first meeting with experienced spelunkers and your first trip into a wild cave.

Clothing

If the "well-dressed" spelunker seems a little sloppy on his way into a cave, you can be sure he will look even worse when he comes out. From head to toe, his outfit has been selected with an

214

appreciation for the fact that caving is both dirty and hard on clothes.

The first essential is proper headgear. A narrow-brimmed hard plastic hat of the type worn by construction works is indispensable. Ceilings are often lower than you might expect, and stalactites are a constant hazard to the head. Besides providing a bracket for holding your headlamp, the hat should have a chin strap to keep it in place when you climb or crawl.

Although no footwear is right for every possible situation underground, good hiking shoes are usually adequate. Most cavers prefer leather boots with tops high enough to support the ankles yet not so high as to cramp the calves. Since leather soles skid on smooth rocks and on mud, select shoes with rubber or composition soles and some sort of tread. Wear two pairs of socks—heavy woolen ones over lightweight inner socks. Rubber boots are desirable in caves that involve wading.

Choice of clothing is more flexible. Remember, however, that mud and clay stain clothing, and projecting rocks can rip even strong material. The best outfit is a moderately loose set of one-piece coveralls. "Levi"-type trousers are too tight for climbing, and separate trousers and jackets tend to snag on rocks in crawlways.

Since most caves are damp and chilly, you normally will need sweaters or sweat shirts. In winter or high in the mountains, insulated underwear is in order. Most spelunkers avoid zippers, since they are too easily clogged by mud. Buttons on pockets will help prevent the loss of small equipment. Light, tough work gloves protect against cuts and possible rope burns, and they will help keep your hands clean in case you plan to use a camera.

HARD HATS AND CARBIDE LAMPS

Light

Always carry several light sources, including spare parts for repairs. Substitutes *must* be immediately available for any light that fails.

The primary light source used by most spelunkers is a carbide headlamp attached securely to a bracket on the hard hat. The lamp casts a broad, suffused light by burning a gas produced by water dripping on pellets of carbide. Extra carbide may be carried in small 35-millimeter film cans, while a plastic bag or other receptacle is necessary for carrying used carbide out of the cave. Besides being a form of litter, spent carbide dropped in the cave can poison cave animals. Conscientious spelunkers bury used carbide outside the cave.

If there is any possibility that oil or other gas-producing material has been dumped in the cave, it is wise to use a headlamp

MAP AND COMPASS

powered by batteries; the open flame of a carbide lamp might ignite the gas. The battery cable makes this type of headlamp somewhat cumbersome, but it has the advantage of providing a long-range spot of light for viewing bats and other cave animals.

If you don't mind carrying your light by hand, you may prefer a waterproof flashlight with spare batteries and bulb. The flashlight also serves as a substitute while you refuel or repair the headlamp. Since it is difficult to climb or crawl by the light of a hand-held lamp, most spelunkers use the flashlight as a supplement, not a primary light source. Remember, too, that cylindrical flashlights tend to roll when you set them down. Rectangular six-volt battery types with handle and lamp attached are better.

It is wise to stow away a couple of large slow-burning "plumbers" candles as well. Besides providing light in an emergency, they can be left burning for hours to mark junctions of passageways and are useful for heating water or food. Don't forget a waterproof container for your matches.

TRAVERSE

Other equipment

A local group probably has mapped the cave; by all means, get the map and study it. A compass is useful whether or not you have a map of the cave. Since passage of time is difficult to estimate underground, you should have a watch. Someone in the party should carry a small first-aid kit for treating cuts and scratches. Any serious injury, however, requires special equipment and trained rescue techniques available only through Civil Defense and cave rescue organizations.

Take along some compact, high-energy food such as candy bars, raisins, or peanuts. Since water flowing through caves may be polluted, carry flat plastic water bottles, both for drinking and for operating your carbide lamp. Never litter; carry empty wrappers, containers, and used flash bulbs out of the cave and keep them until you can dispose of them properly.

As you venture into more difficult caves, you will have to master rope-climbing techniques used by mountaineers. Practice in the sunlight, not in a cave. Take lessons from an experienced rock climber. Besides showing you how to climb, he can advise you in selection of proper equipment.

Procedures

Never enter a cave alone. This is the most important single rule of spelunking. At least two and preferably three or more people should be in every group. A lone caver may be unable to extricate himself after even a minor mishap. Worse still, he has no one to send to the surface for help.

216

Always notify someone on the surface when your party is entering a cave and indicate when you expect to return. Afterward report when all members of your group are out. If you do not return by the specified time, someone can be sent to look for you.

Always study the situation of the cave in relation to the landscape. Unless the cave entrance is on a mountaintop or on the side of a cliff, flooding can be a particularly serious hazard, especially between late winter and early fall. If the entrance is at a low point in the terrain, a tremendous volume of water may pour into the cave during a cloudburst or even an ordinary summer thunderstorm. In areas where flooding is possible, experienced spelunkers usually station a "weather watcher" at the cave entrance.

A cave is no place to prove your daring or your endurance. Fatigue dulls responses and judgment. The result is carelessness—the number-one cause of accidents. Remember that you will often use even more energy coming out of the cave than going in. It is foolish to push on into the cave when you are noticeably tired. Wait until another day to explore that intriguing section just ahead.

If you should get lost, don't panic. Hysterical groping won't get you out of the cave. Keep calm. Conserve your light and your energy. If methodical searching does not lead you back to familiar territory, sit down and wait. Others in your party soon will miss you and begin searching. If everyone is lost, it is just a matter of time until rescuers will be sent into the cave when you fail to return. To keep from getting lost in the first place, watch where you are going and look back frequently to see what formations and other landmarks look like from the opposite direction.

Do not mark cave walls. If you must mark your route, place pieces of luminous tape around formations and remove them on the way out. Carving initials on walls or formations is inexcusable, as is the removal of formations or mineral specimens. They are part of the cave; leave them for others to enjoy. Even broken formations should be left where they are. By removing them, you may encourage other people to take specimens—and they may not be content with broken pieces lying on the floor.

Cave animals are even more easily affected by spelunkers. None are really plentiful, and many kinds could be exterminated by collectors. Watch them, study their habits, and better yet, take pictures. But do not disturb the animals unnecessarily.

Finally there is the matter of courtesy. Most caves are on private land. Ask the owner's permission before entering his land or his cave. Park your car where it is out of his way. Close gates behind you. Walk beside fences, never through the middle of cropland. If the cave entrance is barricaded to keep livestock from falling in, replace the barrier when you leave. Don't litter. Notify the owner when you have finished, and thank him.

RAPPELING

Common Cave Formations

SODA STRAWS are thin-walled hollow tubes about a quarter of an inch in diameter. They grow as water runs through their centers and deposits rings of calcite around their tips. Occasionally they reach lengths of five or six feet.

STALACTITES form as mineral layers are deposited by water flowing over the outsides of soda straws. They usually form when the centers of the soda straws become plugged.

STALAGMITES grow up from the floor where mineral-laden water drips from above, generally beneath stalactites. In contrast to the pointed tips of carrot-shaped stalactites, the tops of stalagmites are blunt and rounded.

COLUMNS are formed when stalagmites meet overhanging stalactites, or when a stalactite grows all the way to the floor or a stalagmite all the way to the ceiling. Water flowing down the sides of the column gradually enlarges it by adding layers of flowstone to the surface.

CAVE GRAPES are irregular clusters of rounded knobs of calcium carbonate. Also known as cauliflower, popcorn, and cave coral, they build up on walls and existing formations in flooded chambers.

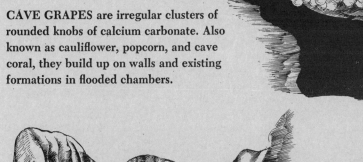

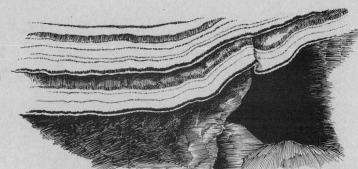

DRAPERIES form where beads of water trickle down the undersides of inclined surfaces. As drop after drop flows along the same irregular course, a thin, often translucent sheet of calcium carbonate gradually extends downward, sometimes for several feet.

BACON FORMATION is drapery with alternating darker and lighter bands. These bands are the result of variations in the mineral content of the trickling water.

RIMSTONE DAMS often create steplike terraces along streams and on cave floors. Though generally only inches high, they occasionally build to heights of several feet and enclose sizable pools of water. Calcite is deposited when the water loses carbon dioxide as it flows over the lips of the dams.

FLOWSTONE forms where films of water flow over walls, floors, and formations, depositing sheets of calcium carbonate like icing.

CAVE PEARLS, generally an inch or less in diameter, form around sand grains and other particles in pools where there is enough agitation from dripping to prevent the pearls from becoming cemented to the bottom. They slowly enlarge as layer upon layer of calcite is deposited.

HELICTITES are small twisted structures that grow from walls, floors, ceilings, and other formations. The contorted forms result from water seeping so slowly through minute central canals that calcite crystals form in irregular positions at the tips.

GYPSUM FLOWERS are found on the walls of many drier caves. They grow from their bases instead of their tips; each petal is pushed outward—sometimes a foot or more—as new crystals form at the bottom.

GYPSUM NEEDLES grow like fragile strands of glass from the sediment of cave floors. Eventually breaking as a result of their own weight, the delicate needles sometimes lie in jumbled heaps.

DOGTOOTH SPAR, large pyramid-shaped crystals of calcium carbonate, forms underwater in flooded chambers. In certain caves in the Black Hills of South Dakota, these crystals are as much as six inches long.

Adaptation: An inherited structural, functional, or behavioral characteristic that improves an organism's chances for survival in a particular *habitat*. *See also* Mutation.

Antenna (plural *antennae*): A feeler; an *appendage*, sensory in function, that occurs in pairs on the heads of crustaceans, insects, and certain other animals.

Appendage: An arm or other limb that branches from an animal's body.

Aquatic: Living in water. Aquatic cave animals include amphipods, isopods, crayfish, *planarians*, fish, and blind salamanders. *See also* Terrestrial.

Arthropods: Animals with jointed legs and hard external skeletons (*exoskeletons*). The group includes insects, *crustaceans*, spiders, millipedes, and several other types of animals commonly found in caves.

Bacteria: Simple, colorless one-cell plants, most of which are unable to manufacture their own food using sunlight. Bacteria are possibly important in caves as synthesizers of food materials from minerals. They are also important as *decomposers*.

Barbels: Fleshy threadlike sensory structures hanging like whiskers near the mouths of certain fish, such as catfish.

Bathybenthic: Of the bottom of the truly deep areas of the sea, where the "rain" of organic material produces a deposit of food.

Bathypelagic: Of the deep sea. Refers to the depths between roughly 3000 feet below the surface and the bottom of the sea. No food accumulates in these waters.

Biological clock: An inherited time-measuring process within a living thing, which governs its responses to certain external events.

Biomass: The total weight of living matter, whether in an entire community, at a particular *trophic level*, or of a particular kind of organism in the community. Thus we may refer to the biomass of a pond community, of herbivores in the pond, or of copepods in the pond.

Biospeleology: The scientific study of cave animal life. A biologist who specializes in this study is called a biospeleologist.

Breakdown: A heap of rock filling all or part of a cave passage after the collapse of part of the walls or ceiling. The term usually refers only to large accumulations of rock.

Carbide lamp: A type of lamp used by miners and cave explorers. It maintains a flame by burning acetylene, a gas produced when water drips on a supply of calcium carbide pellets.

Carnivore: An animal that lives by eating the flesh of other animals. *See also* Herbivore; Insectivore; Omnivore.

Cave: Any natural cavity or series of cavities beneath the surface of the earth. Such cavities are usually classed as caves only if they are large enough to permit entrance by humans. The term is generally synonymous with *cavern* and is commonly applied also to wind- or water-eroded rock cavities.

Cave deposit: An accumulation of material other than *speleothems*, such as charcoal, fossils, clay, silt, gravel, and other floodborne debris.

Cave system: All the cavities and underground passages in a given area, which are now or at one time were interconnected.

Chlorophyll: A group of *pigments* producing the green color of plants; essential to *photosynthesis*.

Climate: The average weather conditions of an area, including temperature, rainfall, humidity, wind, and hours of sunlight, based on records kept for many years.

Column: A pillarlike *speleothem* resulting from the union of a *stalactite* and a *stalagmite* into a single *formation*.

Community: All the plants and animals that live in a particular *habitat* and are bound together by *food chains* and other interrelations.

Competition: The struggle between individuals or groups of living things for common necessities, such as food or living space.

Conservation: The use of natural resources in a way that assures their continuing availability to future generations; the wise use of natural resources.

Constant-temperature zone: The area of a cave where air temperature is unchanging throughout the year and approximates the average annual temperature aboveground. *See also* Zonation.

Consumer: Any living thing that is unable to manufacture food from nonliving substances, but depends instead on the energy stored in other living things. *See also* Carnivore; Decomposers; Food chain; Herbivore; Omnivore; Producers.

Crustaceans: The large class of animals that includes lobsters, crayfish, amphipods, isopods, and many similar forms. Crustaceans typically live in water and have many jointed *appendages*, segmented bodies, and hard *exoskeletons*.

Cupula (plural *cupulae*): A jellylike rod projecting into the water from a *neuromast*, part of a fish's or amphibian's *lateral line system*. Vibrations in the water cause the cupula to move, thus setting off nervous impulses that enable the animal to detect nearby movements in the water.

Decomposers: Living things, chiefly bacteria and fungi, that live by extracting energy from the decaying tissues of dead plants and animals. In the process, they also release simple chemical compounds stored in the dead bodies and make them available once again for use by green plants.

Domepit: A large vertical underground shaft where water flowing down to the *water table* at a lower level has dissolved a cylindrical cavity in the rock.

Drapery: A thin curtainlike *speleothem* that forms where water trickles down an inclined surface.

Ecology: The scientific study of the relationships of living things to one another and to their *environment*. A scientist who studies these relationships is an ecologist.

Embryo: A developing individual before its birth or hatching.

Environment: All the external conditions surrounding a living thing.

Evolution: The process of natural consecutive modification in the inherited makeup of living things; the process by which modern plants and animals have arisen from forms that lived in the past. *See also* Mutation.

Exoskeleton: An external skeleton. The hard body covering or shell of most *invertebrate* animals, including insects, crayfish, and millipedes.

Flowstone: Any mineral deposit that forms on the walls or floor of a cave as a result of water flowing over the surface; often called travertine.

Food chain: A series of plants and animals linked by their food relationships; the passage of energy and materials from *producer* through a succession of *consumers*. Green plants, plant-eating insects, and an insect-eating bat would form a simple food chain. *See also* Food web.

Food pyramid: The normally diminishing number of individuals and amount of organic material produced at each successive level along a *food chain*. The declining productivity at each level results from the constant loss of energy in *metabolism* as the energy passes along the chain. *See also* Trophic level.

Food web: An interlocking system of *food chains*. Since few animals rely on a single food source and since no food source is consumed exclusively by a single *species* of animal, the separate food chains in any natural *community* interlock and form a web.

Formation: A term commonly used for *speleothem*.

Fossil: Any remains or traces of animals or plants that lived in the prehistoric past, whether bone, cast, track, imprint, pollen, or any other evidence of their existence.

Geological map: A map that shows the kinds of rock lying beneath the soil or reaching the surface in a given area. A topographic map shows the contour or elevation lines, and surface features such as watercourses.

Geology: The scientific study of the earth and the rocks that form it. A scientist who specializes in this study is a geologist.

Guano: Excrement, as of bats, crickets, or sea birds. In certain bat caves and on islands colonized by sea birds, guano sometimes accumulates in such vast quantities that it is mined commercially for fertilizer.

Gypsum: Hydrated calcium sulphate, a mineral often appearing as outward-curving petal-like "flowers." The rock, softer and

more soluble than *limestone,* is sometimes massive enough to permit cave formation.

Habitat: The immediate surroundings (living place) of a plant or animal; everything necessary to life in a particular location except the organism itself.

Helictite: A thin, twisting *speleothem* projecting at an angle other than the vertical.

Herbivore: An animal that eats plants, thus making the energy stored in plants available to *carnivores. See also* Carnivore; Insectivore; Omnivore.

Hibernation: A prolonged dormancy or sleeplike state in which animal body processes such as heartbeat and breathing slow down drastically and the animal neither eats nor drinks. Nearly all cold-blooded animals and a few warm-blooded animals hibernate during the winter in cold climates. Extremely large aggregations of bats, crickets, and spiders hibernate in some caves.

Humidity, relative: The ratio, expressed as a percentage, of the amount of water vapor actually present in air of a given temperature as compared with the greatest possible amount of water vapor that could be present in air at that temperature. *See also* Psychrometer.

Infrared light: Light not visible to the human eye, with wavelengths longer than those of visible red light and shorter than those of radio waves.

Insectivore: An animal that feeds on insects. Almost all species of North American bats are insectivores. *See also* Carnivore; Herbivore; Omnivore.

Invertebrate: An animal, such as a planarian, snail, or crayfish, without a backbone. *See also* Vertebrate.

Karst: The typical surface terrain of a *limestone* region, characterized by an abundance of *sinkholes*, disappearing streams, exposed rock outcrops or ledges, and underground caverns. Named for a noted limestone area of northwestern Yugoslavia.

Larva (plural *larvae*): An active immature stage in an animal's life history when its form usually differs from the adult form, such as the grub stage in the development of a beetle or the tadpole stage in the life history of a frog. *See also* Metamorphosis; Pupa.

Lateral line system: A series of sensory organs, usually appearing in a line or series of lines on the sides and heads of fishes and larval amphibians. The system enables the animal to sense vibrations in the water. *See also* Cupula; Neuromast.

Limestone: Sedimentary rock composed primarily of calcium carbonate. It usually originates through the accumulation of calcareous (limy) remains of marine animals. Because limestone is easily dissolved by carbon dioxide in water, caves are more common in limestone than in any other type of rock. Limestone dissolves fastest where the carbon dioxide content is highest—at the surface of the *water table.*

Mammals: The class of animals that includes bats, mice, man, and many others. They typically have a body covering of hair and give birth to living young, which are nursed on milk from the mother's breast.

Marine relict: An animal whose presently extinct ancestors lived in salt water but became adapted to life in fresh water when an area formerly covered by the sea became dry land.

Metabolic rate: The rate at which a living thing transforms food into energy and body tissue. The higher its metabolic rate, the more food it must consume. Most cave animals live at a reduced metabolic rate.

Metabolism: The sum of the chemical activities taking place in the cells of a living thing; the sum of the processes by which a living thing transforms food into energy and living tissue.

Metamorphosis: A change in the form of a living thing as it matures, especially the drastic transformation from a *larva* to an adult. *See also* Pupa.

Microclimate: "Little *climate.*" The environmental conditions, such as temperature, humidity, and air movement, in a very restricted area, such as a sheltered nook in a cave wall.

Microhabitat: A miniature *habitat* within a larger one; a restricted area where environmental conditions differ from those in the surrounding area. A sheltered nook in a cave wall is an example of a microhabitat within the cave.

Mold: A microscopic form of fungus responsible for much food spoilage and, in caves, for conspicuous tufts quickly covering scats, dead insects and bats, and even wooden structures such as ladders.

Mutation: A sudden change in the genetic material of an organism's germ cells, resulting in offspring that possess characteristics markedly different from those of either parent. Mutations generally are harmful but occasionally may improve an organism's chances for survival. *See also* Adaptation; Evolution.

Neoteny: The condition of retaining *larval* form and behavior even as a mature individual. Certain salamanders in particular are neotenic.

Neuromast: One of the individual sense organs that make up the *lateral line systems* of fishes and amphibians. *See also* Cupula.

Omnivore: An animal that habitually eats both plants and animals. *See also* Carnivore; Herbivore; Insectivore.

Organic: Pertaining to anything that is or ever was alive or produced by a living plant or animal. Organic material brought into the cave from outside is virtually the only source of food for cave dwellers.

Paleontologist: A scientist who studies the life of the past by interpreting *fossil* remains of plants and animals.

Photosynthesis: The process by which green plants convert carbon dioxide and water into simple sugar. *Chlorophyll* and sunlight are essential to the series of complex chemical reactions involved in the process.

Pigment: A chemical substance that imparts color to an object by reflecting or transmitting only certain light rays and absorbing all others. For example, a substance that absorbs all but green rays appears green. An object that contains no pigment, on the other hand, appears white because it reflects all light rays and absorbs none. Many troglobites have lost all their pigment.

Planarian: A flatworm. A relatively simple wormlike animal with a flattened ribbonlike body, a distinct head end, and a mouth located more or less centrally on the underside of the body.

Pleistocene: Of or pertaining to the most recent period in the earth's history, roughly the past one million years. The period includes at least four major retreats and advances of continental glaciers.

Pollution: The fouling of water or air with sewage, industrial wastes, or other contaminants, making them unfit to support many forms of life. Pollution can be especially serious underground where extensive networks of passages spread contaminating materials for long distances.

Preadapted: Possessing *adaptations* that would contribute to survival in a *habitat* other than the immediate one because of similarities in living conditions in the two habitats. Insects that live in leaf litter on the forest floor, for example, may be preadapted to cave life.

Predator: An animal that lives by capturing other animals for food. *See also* Prey.

Prey: A living animal that is captured for food by another animal. *See also* Predator.

Producers: Green plants, the basic link in any *food chain;* by means of *photosynthesis,* green plants manufacture the food on which all other living things ultimately depend. They are available in the cave *community* only in the *twilight zone,* or as debris that falls or washes in. A few types of bacteria also manufacture food from nonliving substances and therefore serve as producers in some cave communities. *See also* Consumer.

Psychrometer: An instrument used for measuring *relative humidity.* The simplest psychrometers consist of two thermometers mounted on a rotating frame. One thermometer's bulb is kept moist, the other dry. By comparing the readings of the two thermometers after they have been whirled in the air, one can determine the relative humidity.

Pupa (plural *pupae*)**:** The inactive stage in the life history of certain insects during which the *larva* undergoes a gradual reorganization of its tissues in the process of becoming an adult. *See also* Metamorphosis.

Scats: Animal droppings, an important source of food in caves.

Scavenger: An animal that eats the dead remains and wastes of other animals and plants. *See also* Predator.

Sinkhole: A surface depression in cave country. A sinkhole is produced when the roof of a cave collapses or when *limestone* rock underlying the soil is slowly dissolved by water.

Soda-straw stalactite: A thin-walled tubular *stalactite* that elongates as minerals are deposited at the tip by water dripping through its hollow interior. All stalactites begin their growth as soda straws.

Sonar: A system for detecting obstacles by emitting sound and intercepting and interpreting echoes that bounce back. It is used by bats and also by oilbirds and some swiftlets when they fly in the darkness of caves.

Species (singular or plural): A group of plants or animals whose members breed naturally only with each other and resemble each other more closely than they resemble members of any similar group.

Speleologist: A person who studies caves in any of their scientific aspects. *See also* Spelunker.

Speleothem: A general term for any mineral deposit or formation found in caves, such as *stalactites, stalagmites,* or *gypsum* flowers.

Spelunker: A person who explores caves as a hobby or for recreation. *See also* Speleologist.

Stalactite: An iciclelike deposit of calcium carbonate which grows downward from the ceiling of a cave. *See also* Speleothem; Stalagmite.

Stalagmite: A deposit of calcium carbonate which builds upward from a cave floor as a result of water dripping from above. *See also* Speleothem; Stalactite.

Terrestrial: Living on land. Terrestrial cave animals include blind beetles, millipedes, spiders, and crickets. *See also* Aquatic.

Troglobite: "Cave dweller." An animal that lives in caves and nowhere else.

Troglophile: "Cave lover." An animal that can complete its life cycle in caves but may also do so in suitable *habitats* outside caves.

Trogloxene: "Cave visitor." An animal that habitually enters caves but must return periodically to the surface for certain of its living requirements, usually food.

Trophic levels: Feeding levels in a *food chain,* such as *producers, herbivores,* and so on. Most food chains include a maximum of four or five trophic levels.

Twilight zone: The area of a cave where light penetrating through the entrance is sufficient to permit human vision. *See also* Zonation.

Variable-temperature zone: The area of a cave where air temperature fluctuates with the seasons. *See also* Zonation.

Vertebrate: An animal with a backbone. The group includes fishes, amphibians, reptiles, birds, and mammals. Some amphibians and fishes live permanently in caves. *See also* Invertebrate.

Water table: The upper level of the underground reservoir of water; the level below which the soil and all cracks and channels in the rocks are saturated.

Zonation: The organization of a *habitat* into a more or less orderly series of distinctive plant and animal associations as a result of variations in environmental conditions. Zones in a cave are the *twilight zone,* the *variable-temperature zone,* and the *constant-temperature zone.*

SPELEOLOGY

CULLINGFORD, C. H. D. (Editor). *British Caving: An Introduction to Speleology.* Routledge, 1962.

LAWRENCE, JOE, JR., and ROGER W. BRUCKER. *The Caves Beyond.* Funk & Wagnalls, 1955.

MOORE, GEORGE W., and BROTHER G. NICHOLAS. *Speleology: The Study of Caves.* Heath, 1964.

CAVES AND CAVE AREAS

BAILEY, VERNON. *Animal Life of the Carlsbad Caverns.* Williams & Wilkins, 1928.

BARR, THOMAS C., JR. *Caves of Tennessee.* Tennessee Division of Geology, 1961.

BRETZ, J. HARLEN. *Caves of Missouri.* Missouri Division of Geological Survey and Water Resources, 1956.

BRETZ, J. HARLEN, and S. E. HARRIS, JR. *Caves of Illinois.* Illinois State Geological Survey, 1961.

DAVIES, WILLIAM E. *Caverns of West Virginia.* West Virginia Geological Survey, 1949.

DAVIES, WILLIAM E. *The Caves of Maryland.* Maryland Department of Geology, Mines and Water Resources, 1950.

DOUGLAS, H. H. *Caves of Virginia.* Virginia Cave Survey, 1964.

HALLIDAY, WILLIAM R. *Caves of Washington.* Washington State Department of Conservation, 1963.

JACKSON, GEORGE F. *Wyandotte Cave.* Livingston Publishing Co., 1953.

MOHR, CHARLES E. (Editor). *The Caves of Texas.* National Speleological Society, 1948.

PERRY, CLAY. *New England's Buried Treasure.* Stephen Daye Press, 1946.

PERRY, CLAY. *Underground Empire.* Stephen Daye Press, 1948.

POWELL, RICHARD L. *Caves of Indiana.* Indiana Department of Conservation, 1961.

SPANGLE, PAUL F. (Editor). *A Guidebook to Carlsbad National Park.* National Speleological Society, 1960.

BIOLOGY AND ECOLOGY

BUCHSBAUM, RALPH, and MILDRED BUCHSBAUM. *Basic Ecology.* Boxwood Press, 1957.

EIGENMANN, CARL H. *Cave Vertebrates in North America: A Study in Degenerative Evolution.* Carnegie Institution of Washington, 1909.

FARB, PETER, and THE EDITORS OF LIFE. *Ecology.* Time, Inc., 1963.

ODUM, EUGENE P., and HOWARD T. ODUM. *Fundamentals of Ecology.* Saunders, 1959.

STORER, JOHN H. *The Web of Life.* Devin-Adair, 1960.

VANDEL, A. *Biospeleology: The Biology of Cavernicolous Animals.* Pergamon Press, 1965.

PLANTS

CHRISTENSEN, CLYDE. *The Molds and Man.* University of Minnesota Press, 1951.

WILSON, CARL L., and WALTER E. LOOMIS. *Botany.* Holt, 1962.

ANIMALS

BISHOP, SHERMAN C. *Handbook of Salamanders.* Cornell University Press, 1947.

BUCHSBAUM, RALPH. *Animals without Backbones.* University of Chicago Press, 1948.

GRIFFIN, DONALD R. *Echoes of Bats and Men.* Anchor Books, 1959.

GRIFFIN, DONALD R. *Listening in the Dark.* Yale University Press, 1958.

PALMER, RALPH S. *The Mammal Guide.* Doubleday, 1954.

PETERSON, RUSSELL. *Silently, By Night.* McGraw-Hill, 1964.

SCHWARTZ, CHARLES W., and ELIZABETH R. SCHWARTZ. *The Wild Mammals of Missouri.* University of Missouri Press, 1959.

EVOLUTION

BURNETT, A., and T. EISNER. *Adaptation: A Case Study.* Holt, 1964.

COLBERT, EDWIN H. *Evolution of the Vertebrates.* Wiley, 1955.

MOORE, RUTH, and THE EDITORS OF LIFE. *Evolution.* Time, Inc., 1962.

SIMPSON, GEORGE GAYLORD. *The Major Features of Evolution.* Columbia University Press, 1953.

SIMPSON, GEORGE GAYLORD. *The Meaning of Evolution.* Yale University Press, 1950.

GEOLOGY

DURY, G. *The Face of the Earth.* Penguin, 1959.

LONGWELL, CHESTER R., and RICHARD F. FLINT. *Introduction to Physical Geology.* Wiley, 1962.

THORNBURY, W. D. *Principles of Geomorphology.* Wiley, 1954.

GENERAL READING

CASTERET, NORBERT. *The Darkness under the Earth.* Holt, 1954.

COON, C. S. *The Seven Caves: Archaeological Explorations in the Middle East.* Knopf, 1957.

DOUGLAS, JOHN SCOTT. *Caves of Mystery.* Dodd, Mead, 1965.

FOLSOM, FRANKLIN. *Exploring American Caves.* Collier Books, 1962.

HALLIDAY, WILLIAM R. *Depths of the Earth.* Harper, 1966.

LÜBKE, ANTON. *The World of Caves.* Coward-McCann, 1962.

MOHR, CHARLES E., and HOWARD N. SLOANE. *Celebrated American Caves.* Rutgers University Press, 1955.

PINNEY, ROY. *The Complete Book of Cave Exploration.* Coward-McCann, 1962.

SCHWARTZ, DOUGLAS W. *Prehistoric Man in Mammoth Cave.* Eastern National Park and Monument Association, 1965.

TAZIEFF, HAROUN. *Caves of Adventure.* Harper, 1953.

THRAILKILL, JOHN. *Introduction to Caving.* Highlander Publishing Co., 1962.

227

Illustration Credits and Acknowledgments

COVER: Formations one thousand feet from the entrance of McClung Cave, West Virginia, Lyle G. Conrad

UNCAPTIONED PHOTOGRAPHS: 8–9: Entrance of Russell Cave, Alabama, Lyle G. Conrad 58–59: Flight of Mexican free-tailed bats, Ney Cave, Texas, Robert W. Mitchell 98–99: Cave salamander (*Haideotriton wallacei*) collected in Gerard's Cave, Florida, Robert S. Simmons 148–149: Stalactites and stalagmites in Caverns of Sonora, Texas, Mills Tandy

ALL OTHER ILLUSTRATIONS: 10: Teresa Zabala 11–13: Earl H. Geil 14: Lyle G. Conrad; Peter G. Sanchez 15: Mills Tandy; Peter Lindsley 16: Peter G. Sanchez 17: Charles E. Mohr 18–19: Earl Neller 20: Thomas C. Barr; William J. Steffenson 21: Charles E. Mohr; Francis E. Abernathy 22: John Gerard from National Audubon Society 23: Charles E. Mohr 24: Patricia Henrichs 25: Charles E. Mohr 26–28: Robert W. Mitchell 29: Charles E. Mohr 30: J. E. Cooper 31: William T. Austin from Cave Research Foundation; James L. Papadakis 32: William T. Austin from Cave Research Foundation 33: Charles E. Mohr 34–35: Earl Neller 36: Charles E. Mohr 37–39: William W. Varnedoe 40–43: Felix Cooper 44: Charles E. Mohr 45: Ansel Adams 46–47: Bruce Roberts from Rapho-Guillumette 48–49: A. Y. Owen, courtesy of Curtis Publishing Company 50: Bruce Roberts from Rapho-Guillumette 51: Charles E. Mohr 52–53: Patricia Henrichs 54–55: Charles Fracé 56: Charles E. Mohr 60: Robert W. Mitchell 61: Charles E. Mohr 62: Lyle G. Conrad; Thomas C. Barr 63: Robert W. Mitchell 64: Charles E. Mohr 65–67: Graphic Arts International 68–69: Charles E. Mohr 70: Peter G. Sanchez 71: Robert W. Mitchell 72: Peter G. Sanchez 73: Mills Tandy 74–75: Peter G. Sanchez 76–79: Robert W. Mitchell 80: Russell Gurnee 81: Charles E. Mohr 82–87: Robert W. Mitchell 88: Charles E. Mohr 89: Graphic Arts International 90–91: Art Dobson 92: Graphic Arts International 93: Charles E. Mohr 94–95: Charles Fracé 96: Charles E. Mohr 100: Robert W. Mitchell 101: Larry West from National Audubon Society 102: John Gerard 103: Charles E. Mohr 104: Charles Fracé 105–109: Edward S. Ross 110: Graphic Arts International 111: Robert W. Mitchell 112–113: Frederic A. Webster; Charles Fracé 114–116: Charles E. Mohr 118–119: Charles Fracé 120: R. L. Powell 121: Robert W. Mitchell 122: Robert W. Mitchell; Graphic Arts International 123: Robert W. Mitchell; Graphic Arts International 124: Charles E. Mohr 125: W. Ray Scott, National Park Concessions, Inc. 126–127: Charles E. Mohr 128: Felix Cooper 130–131: Robert W. Mitchell 132–133: Mills Tandy 134–136: Charles E. Mohr 137: Graphic Arts International 138: Patricia Henrichs 139: Charles E. Mohr; Robert S. Simmons 141: Charles E. Mohr 142–143: Graphic Arts International 144: William H. Amos 145: Charles E. Mohr 146: Robert W. Mitchell 150: Howard N. Sloane 151–152: Charles E. Mohr 153: Patricia Henrichs 154: Graphic Arts International 155: Charles Fracé; Charles E. Mohr 156: Charles E. Mohr 157: St. Louis Post Dispatch; Charles E. Mohr 158: Charles E. Mohr; Graphic Arts International 159: Charles E. Mohr 160: Richard Raber 161: John Guilday, Carnegie Museum 162: Patricia Henrichs 163: George Porter from National Audubon Society 164–165: Patricia Henrichs 166–167: Charles Fracé 168–169: Mills Tandy 170–171: Robert W. Mitchell 173: William T. Austin from Cave Research Foundation 174: Robert W. Mitchell 175: William Taylor 176–177: Graphic Arts International 178: Mills Tandy 179: Robert W. Mitchell 180: A. Y. Owen, courtesy of Time, Inc. 181–183: Robert W. Mitchell 184: Charles E. Mohr 185–187: Robert W. Mitchell 188–189: Patricia Henrichs 190: Richard Peterson 192–193: William T. Austin from Cave Research Foundation; W. Ray Scott, National Park Concessions, Inc. 194–195: Thomas C. Barr 196–197: William T. Austin from Cave Research Foundation 198: Peter Lindsley 201–209: Charles Fracé 214–217: Charles Fracé 218–220: Patricia Henrichs

PHOTO EDITOR: ROBERT J. WOODWARD

ACKNOWLEDGMENTS: *The publisher is particularly grateful to Robert W. Mitchell of Texas Technological College for assisting the editors in selecting photographs, for providing many of the photographs, and for offering valuable criticism of the text. The publisher also wishes to thank the following individuals: John Guilday of the Carnegie Museum, Brother G. Nicholas, F.S.C., of La Salle College, and Donald R. Griffin of Rockefeller University for reading and criticizing portions of the manuscript; Gerald Frederick, Richard R. Anderson, and Russell Gurnee of the National Speleological Society and all members of the society who so willingly submitted their photographs to the editors for consideration; C. Gordon Fredine, William Perry, Wayne W. Bryant, M. Woodbridge Williams, and Philip F. Van Cleave of the National Park Service for their helpful criticism of the manuscript and for their assistance in locating photographs.*

Index

[Page numbers in boldface type indicate references to illustrations.]

231